Copying, Duplication
and Microfilm

Copying, Duplication and Microfilm

systems and equipment for use
in business and administration

HARRY T. CHAMBERS

London

BUSINESS BOOKS LIMITED

First published 1970
Second edition 1972

ISBN 0220 66843 4

A
686.4

This book has been set in 10 on 12 pt Pilgrim
in England by Clarke, Doble & Brendon Ltd., Plymouth
for the publishers, Business Books Limited
(registered office: 180 Fleet Street, London, E.C.4),
publishing offices: Mercury House, Waterloo Road, London SE1 8UL

MADE AND PRINTED IN GREAT BRITAIN

Contents

List of figures

List of plates

(between pages 86 and 102)

Acknowledgements

The author wishes to acknowledge the help of many friends concerned in the manufacture and use of copying and duplicating equipment. In particular thanks are due to Rank Xerox Ltd for supplying illustrations covering xerographic equipment; Ofrex Ltd for illustrations of thermographic equipment and Bell and Howell Ltd for photographs of microfilm equipment. Acknowledgements are also made to the editor and publishers of *Business Administration* for permission to use the illustration showing invoicing by the copy system which first appeared in that magazine.

The author would further like to acknowledge the help he has received from The National Cash Register Co Ltd with regard to PCMI discussed in the second edition. Office Equipment (John Dale) Ltd for clarifying some of the problems associated with microfilm storage and retrieval and both Kodak and Bell and Howell for demonstrating various of their systems.

Introduction to the second edition

In the introduction to the first edition I started by asking why we make copies. My short answer then was to convey information in a visible and permanent form.

The reason is still the same. We make copies to shortcircuit the long delays which would happen in modern business if the original documents were to be passed up and down the lines of communication. Asked to bring this book up to date it soon became obvious that little has happened to change the basic needs for making copies or in the management structure which gives rise to the need to communicate with several people simultaneously.

Even then I pointed out that the difference between copying and duplicating was no longer so important in practice as it had been. This is even more true today than when I first pointed it out. The technical difference remains but for the user it is frequently a choice between one process or another regardless of which is technically copying or duplicating. Detail changes there have been in plenty; both copying and duplicating equipment has been improved. Whereas when I first sat down to write the first edition electrostatic copiers were rapidly coming to dominate the copying scene, this process is now virtually complete except for the smallest applications.

Some people in the industry grumble that we are in an electrostatic age and that other equipment would do the job better if it did not suffer from an old fashioned image. There may be a germ of truth in this so far as specific applications are concerned. Nevertheless I am willing to stick my neck out and say that a more probable outcome is the development of electrostatics to a point where inkless printing is a reality—at a printing price. Today, too, most users are more cost conscious than they were even two years ago.

However, what has emerged on the business administration

scene is microfilm as a viable information storage and retrieval system. It has been there for years in the background, but today it is being taken seriously. Consequently this second edition is very different from the first. In the comparatively short space of time between the first and second editions everything I then had to say about microfilm has become outdated. In this edition I have set out to explain what microfilm can do, how it does it and how to set about finding out whether microfilm can help to solve your particular information storage and distribution problems. Microfilm has many advantages in business as it has in the academic world. These too I have discussed; after all hard copy communication is the same in a university administration as in a business administration. Well, nearly so. University administration was once described to me as one whacking great personnel department where most of the personnel are the customers. There is a good deal in that. I hope teaching staff will forgive me for saying that they are in the business of 'selling' information.

Whatever way you look at it a student goes to university to learn and research and for that he needs background information *ad infinitum*. Whether he gets it from printed books, duplicated or copied sheets or from microfilm is immaterial to the student.

At the same time similar criteria and techniques apply in business administration where learning in the broadest sense goes on until retirement. More and more efficient storage and distribution of information contributes to efficiency in modern economic life. And that is where we came in—copying, duplication and microfilm all have their part to play.

Copying and duplicating terms

AUTOPOSITIVE PHOTOSTABILIZATION   The same equipment can be used to produce either negative (usually paper) or a direct positive black or white copy. Chemically the process is nearer to ordinary photography than other copying processes and photographic prints can be made from negatives.

CYLINDER   This word is used more in duplicating than in copying. It usually means a hollow cylinder and is normally prefixed by a word which is descriptive of its use. For example, 'ink cylinder' on a stencil machine or 'blanket cylinder' on an offset duplicator. Small solid cylinders are usually referred to as rollers. In duplicating the following are usually met with :

1 MASTER CYLINDER   This is the cylinder around which the master is wrapped in spirit or offset duplicating. In the latter process it is more usual among printers to refer to the master as a plate and hence to the master cylinder as a plate cylinder. The term 'master' is an American import which came to this country with offset duplicators made by companies with American associations.

2 INK CYLINDER   This refers to the cylinder of a single cylinder stencil duplicator. On such machines the ink supply is actually in the cylinder. The outer casing of this cylinder is perforated to allow the ink to pass through to an absorbent 'blanket' or pad.
   Stencil duplicators which have the ink transferred to the outer surface of the cylinder by means of rollers are said to have solid cylinders, although this refers to the walls of the cylinder, the inside usually being hollow for the sake of lightness.

3 BLANKET CYLINDER   Found on offset duplicators where it rotates in contact with the master cylinder. A rubber blanket is wrapped round the blanket cylinder on to which the image transfers from the master, i.e. 'offsets'.

4 IMPRESSION CYLINDER   This is the cylinder which, on most offset duplicators, holds the paper against the blanket cylinder while it is passing through the machine.

5 ROLLERS   These are cylinders of comparatively small diameter used in some stencil and all offset duplicators to convey the ink to the cylinder around which the stencil or plate is wrapped. In offset machines there are also damping rollers. In photocopying a system of rollers is frequently used to convey the paper through the developer unit.

DEVELOPER UNIT   This is the unit in which the copy is developed after the original has been exposed together with the copy paper. The same term is used regardless of the process, even where 'developing' is strictly speaking a misnomer for some other reaction which has the same effect, i.e. to make the copy visible and reasonably permanent.

DIAZO   The copy paper is coated with diazonium salts which have the property of being destroyed by ultraviolet light. On exposure the image blocks the light in the image area. The barely readable yellow copy is then left. This is developed by being combined with an azo dye, also combined in the original coating.

In drawing office equipment an alkali, usually ammonia, is used to release the dye but in office copiers a dye which is released by heat is used.

DIFFUSION TRANSFER   Exposure is by the reflex method and a special silver halide negative paper is used. In the image area the silver salts are loosened by the 'developer' and when the negative is squeezed through the rollers together with a piece of positive paper some of the image is transferred to the latter. Three or four copies are produced with normal quality materials. Today the process is used more to produce size-for-size metal small-offset plates than for document copying.

DUSTING POWDER   This can be used on either small-offset or stencil machines to dry the work as it comes off the machine. For this purpose an automatic powder spray is necessary, known as a 'puffer'. It is available as an optional extra for most small-offset and electric stencil duplicators.

ELECTRICAL SUPPLIES, SAFETY Most copiers and duplicators work from the normal single phase mains. It would be an exceptional office machine which required a three phase supply and such a machine would certainly be installed by a service engineer. Almost without exception a copier or duplicator can be safely connected to a 13-amp socket. The nearest equivalent on the old round pin system is 15 amps. On the standard three square pin 13-amp plug the earth pin is the one at the front of the plug, slightly larger than the other two. This must always be connected to the earth wire of three core cable. In the standard colour-coding system this is green and yellow, but on some imported machines other colours are found from time to time. In case of doubt consult the manufacturer. The 'live' wire must be connected to the fused pin and the neutral to the pin opposite the earthed pin. Again on imported machines the colour may vary from the standard practice which is to provide a brown insulation for the live and blue for the neutral. It cannot be too strongly emphasized that the plug must be correctly connected. Sufficient of the insulation must be stripped back to form a sound connection but bare wires must never touch each other. Only wire a plug yourself if you are sure that you know exactly what you are doing. Faulty wiring can cause either a fire or a short. This latter should simply blow a fuse but it has been known to cause an excessive voltage to be fed to the equipment which can damage it. The correct fuse should be used to suit the type of equipment; it will not protect it if a fuse of a higher rating is put in the plug.

All cables should be sound and if connections are necessary between the apparatus and the socket these should be made with approved connectors. Wrapping insulating tape round the bared ends of the wires which have been twisted together to form a connection is another dangerous practice. Other things to avoid are putting cables under carpets or rugs or leaving them trailing in places where people walk so that someone is likely to trip over them. Most dangerous of all is sticking pins through the insulation of a cable to fix it to a cabinet—I once actually saw this done.

ELECTROSTATIC There are two electrostatic processes. One is the xerographic process developed by the Xerox Corporation, USA (formerly Halide Corporation) and pioneered in Britain by Rank Xerox. The process copies onto plain paper. All others which copy

B

onto coated paper are derived from the Electrofax patents developed by the Radio Corporation of America, who have licensed machine manufacturers in various countries to use their patents.

The basic principle depends on using materials which hold an electrical charge in the dark and disperse it when exposed to light. The image on the original shields that area from light so that the charge is held but dispersed over the rest of the surface when original and copy paper are exposed together. Powder charged with an opposite polarity is then attracted to the still charged image area. This powder is then fused to the surface by heat to make a readable copy.

This is now the most commonly used form of copying and equipment is available to make from 6 to 60 copies a minute, according to the model.

EXPOSURE UNIT   This is the part of a copying machine which houses the light source. Exposure units may be either flatbed or rotary. (See under appropriate term.) They are sometimes available as separate units.

FACE   The word is used to denote the design of a type. In duplicating it will be met with most frequently in connection with typewriters, modern machines being available with a number of different typefaces. A face which has strokes at the top and bottom of the letters is called a 'serif' face, and one which does not is called 'sans serif'. These finishing strokes are usually so small that you only become aware of them when you change to a sans serif face and notice subconsciously that they are no longer there.

FINISHING   To a printer this covers all the processes which take place after the work comes off the press. The word is often used in the same context in duplicating. Finishing processes include cutting, trimming, gathering and collating, as well as forms of binding and laminating. All these can be performed in the office on duplicated work.

1 For duplicating, as opposed to internal printing, a simple hand guillotine is usually sufficient. Usually about 20 or 30 sheets of average thickness can be cut at a time. Such guillotines are inexpensive and are handy for a number of jobs. Do ensure that the

one you have is equipped with a guard which complies with the requirements of the Offices, Shops and Railway Premises Act, 1963.
2 There are a large number of collators available both semi-automatic and fully automatic. Both may be either vertical or horizontal. There is also a third type which works on the principle of a revolving drum. Even the simplest semi-automatic models enable papers to be sorted faster than is possible by hand.

The number of stations and hence the number of pages they can sort at one time is a governing factor.

Some duplicators and copier/duplicators can have collators attached to them so that the papers are sorted as they come off the machine, of maximum use when the duplicator is used for systems work.

3 Binding may be either permanent (adhesive) or loose leaf. For permanent binding there are several pieces of equipment available which can be used in the office. However, the secret of easy secure binding is more in the materials than the equipment. The equipment merely holds the work in position while it is being set by an infrared heater. The same equipment can be used for padding, which is simply a form of semi-permanent adhesive binding for situations where you want to tear off a sheet at a time.

Loose-leaf binding is usually achieved in the office by using a plastic spine in which a number of teeth have been cut. The whole spine is then formed into a tube with the end of the teeth resting inside the spine. The papers are first drilled and then inserted over the teeth. The equipment, therefore, consists of a paper drill and a device to insert the 'teeth' of the spine through the holes drilled in it. Other types of semi-permanent loose-leaf binding are available, the simplest of which consists of a split tube made of plastic, friction keeping the sheets together once the 'spine' has been drawn over one edge of the paper. The only snag is that the papers will not open flat.

4 Laminating, as the term is understood here, refers to laminating a clear film to something which has been duplicated or printed. Laminating films can be applied to paper but it is more usual to apply them to thin card. The most common use for laminators in a duplicating unit is to laminate small articles such as works identity cards to make them more durable. Several kinds of laminators are available, of which the heat type is the cleanest and most practical for office use. Where there is only an occasional need for laminat-

ing, a heat transfer copier can be used although the materials cost is rather high when compared to the materials used in a machine made for the purpose.

Laminating covers to improve their appearance and durability is best left to an outside supplier unless an internal print department with suitable equipment is available within the organization.

FIRE PRECAUTIONS   Duplicating implies the presence of a fair quantity of paper, and other inflammable materials are also used in both copying and duplicating. The following precautions are necessary:

1   Ensure that you have adequate fire extinguishers available and that they are of the right type to deal with paper fires.
2   If smoking is allowed provide adequate ashtrays and insist that they are used.
3   Do not allow smoking in a stationery store.
4   Make sure everyone knows what to do should a fire occur. Even a serious and rapidly spreading fire can be dealt with successfully provided everyone knows what to do.
5   Each fire brigade has a fire protection officer. Call him in and take his advice.
6   Always keep regular and emergency exits free and ensure that any passages leading to them are uncluttered.

FLATBED   This refers to the type of copier where the original is placed on a glass or other translucent platen while it is being copied. This surface is usually covered by a lid or curtain to exclude extraneous light while the original is being copied.

With flatbed machines it is the size of the platen which governs the size of original which can be copied. This is one of those obvious factors which are sometimes overlooked when ordering new equipment.

There are also some flatbed duplicators on which the master is positioned in a flat position and the inking, damping rollers and copy paper brought into contact with it either mechanically or by hand. These are slow compared to rotary machines and although cheap are seldom seen in offices today.

FLUIDS   This is an euphemism for liquid chemicals used in most photocopying processes, it being considered by some copier manufacturers that the word 'chemical' is associated with something too complex for office use.

GELATINE TRANSFER In practice much the same as diffusion transfer. Marketed only by Kodak the trade name being Verifax. The negative, called a matrix, has a gelatine based coating containing a dye-forming chemical. The positive (final copy) paper is uncoated.

GRAMS PER SQUARE METRE This is the modern way of expressing the weight of paper. From 1970 it should have been used exclusively in preference to older methods, and is always used in conjunction with International Paper Sizes.

In the trade, paper is sold by weight and this method gives a straight comparison between the cost of using paper of different substances. Hence, the heavier the substance the more a given number of sheets of the same size and quality will cost. Conversion tables are available to convert the traditional ways of expressing paper substance into grams per square metre but this need not bother the ordinary duplicator user. (See also Paper, weight.)

HEAT TRANSFER There are several versions, as follows:

1 INFRADEX REFLEX This relies on a paper which is sensitive to heat. Carbon or metallic content in the image on the original causes the image area to heat more rapidly than the rest of the copy paper. This is sufficient to cause a chemical reaction in the copy paper which blackens the image area. Very fast copies in 3 to 5 seconds.

2 DUAL SPECTRUM This involves exposure by light and development by heat. Chemically complex but the practical result is to enable any original to be copied. But it is a two-step process. Marketed only by the 3M company.

3 INFRARED TRANSFER The copy 'negative' or intermediate is coated with wax. Also relies on heat differential to transfer part of the coating only in the image area. Metallic or carbon content necessary in image on original. Mostly used for making spirit duplication masters, more recently for stencils.

IMPRESSION CYLINDER (See Cylinders.)

INDIRECT IMAGING OF PLATES Various methods of imaging plates or, as a printer would say, making plates, are described in this book. The important thing to remember is that the equipment and materials for each method can be used with that method only. They are not interchangeable.

INK Different types of ink are used in stencil duplicators and offset duplicators:

   1 STENCIL DUPLICATOR INKS In stencil duplicating ink has to be squeezed through the base fibre of the stencil wherever 'holes' have been cut in the coating. For this reason stencil duplicators need a fair quantity of ink and to prevent this drying out in the machine a type of ink which does not dry out in the atmosphere is used. Some stencil inks are sold in the form of paste and some in the form of liquid. Both consist of a pigment mixed with a colourless oil. Thick inks give a sharper impression but penetrate the copy paper more slowly. Hence you may not be able to run the duplicator at its fastest speed when using a thick ink. In practice, unless you are expert, it is best to keep to the ink supplied by the duplicator manufacturer as this has been formulated for use on the machine in question.

   2 OFFSET INKS Offset inks dry by a mixture of evaporation or oxidation and by absorption into the copy paper. They are normally sold as paste, usually in 1 pound or larger tins. However, for office offset duplicators they are also available ready mixed in tubes. Although this is a rather more expensive way of buying them it is convenient and for the new operator obviates the possibility of misjudging the consistency. Most printing inks sold as paste require mixing with a reducing medium. If this is not so it will say so on the tin. Offset printing ink must be well beaten into a workable condition until it spreads easily on the ink slab with slight pressure of the knife. A special flexible knife rather like a palette knife is used.
       Both stencil and offset inks are available in a range of colours. The important thing to remember is not to try to use offset ink in a stencil machine or vice versa. With both types

of machine it is advisable to use ink supplied by the machine manufacturer until you are fairly expert. Then, in the case of small-offset, you can if you wish buy ink from a printing ink manufacturer. This is often less expensive than the ink sold by machine manufacturers who feel that they are providing an extra service in monitoring the quality of the ink supplied under their labels.

LATENT IMAGE   The image left on the copy paper after exposure but before development, which is invisible in many processes, is known as a 'latent image'. This image needs to be 'developed' before it is of any use as a copy. In electrostatic copying, it is in the form of an electrical charge which would quickly disperse if the paper were to be removed from the machine at this stage.

LIGHT SOURCE   Many kinds of artificial light sources are used in copying. Tungsten bulbs similar to motorcar headlamps and fluorescent tubes are the most common. It is important that the light source should be of the correct type and value. Hence it needs to work at its rated value. If it is wearing out and a replacement cannot immediately be made, compensation can be made by increasing the length of exposure.

MASTER   Masters are the intermediate copy used in duplicating from which the final copies are made. Modern developments have somewhat confused the use of this term, and the following may be helpful in clarifying the situation :
1 In offset work printers always refer to the master as a 'plate'. 'Master' is an American term synonymous with plate. It is used frequently in office and in-plant offset duplicating particularly when referring to a paper plate.
2 On copier/duplicators the master is the original, no separate master being required, except that when using adherography the original must be prepared on a special coated paper. You can think of this either as the original or a master depending on whether the information has to be transcribed on to it to enable it to be duplicated.
3 In stencil duplicating the master is usually known as the stencil. This is simply a matter of habit.
4 In diazo work a translucent copy made for the purpose of

making further copies by the diazo process is usually referred to as a diazo master.

5 The term 'master' is used in a straightforward way in spirit duplicating to mean the mirror copy from which the final copies are made. This applies whether the master is made mechanically or by a duplicating process.

NEGATIVE Originally this word was used to describe a photographic negative. That is to say, it is a copy on which the black and white areas were reversed, when compared with the original. Today the word is used loosely to mean any intermediate copy necessary between the original and the production of the final copy. In some processes it is the tones which are reversed, as in photography, whereas in other processes the 'negative' may be a mirror image of the original. In either case it performs the same function in 'copying' as the 'master' does in duplicating.

NON-REPRODUCING This term refers to either colours or inks which have special characteristics which render them blind to the process in which they are being used. Used on originals such as forms they allow guidelines and other information to be put down on the original which is not required on the copy.

PAPER

1 OFFSET DUPLICATION Offset duplicators today are capable of printing a wide range of papers. Cartridge, banks and bonds all print well. Some art papers and newsprint are more difficult and should only be attempted by an experienced operator. However, today there are some art papers on the market which have been made for offset printing and when ordering art paper the fact that it is to be printed offset should be stated. In practice an offset duplicator used in an office should be used with one or more of the range of papers made for small-offset duplicators. Internal print departments employing experienced operators can use almost any type of paper and also print a number of other materials. Do not, however, attempt to print gummed backed paper unless you know exactly what you are doing. In offset work the paper is damped on the machine.

2 PHOTOCOPIERS It is the process which governs the coating on the paper used for both intermediate (negative) and final copies. For example, in a diffusion transfer copier you must use diffusion transfer materials. You do not necessarily need to use the materials supplied by the company which sold you the machine. However, in the case of the machine producing unsatisfactory results they will undoubtedly blame the materials if they have not themselves supplied them. This is because copier manufacturers make most of their profits from selling the materials rather than the machines.

There are, as you can see from this book, a number of processes on the market where the materials and equipment are manufactured by one company. When you choose a copier using one of these processes you are in the hands of the manufacturer so far as both price and continuity of supplies are concerned.

With old and well established manufacturers this need not cause concern, but if at any time you are offered a copier which uses an exclusive process by an unknown manufacturer it is as well to be assured of his intention to establish a lasting business.

In addition there are some copying processes where the copies are produced on plain paper. Theoretically any suitable paper can be used of a grade similar to that supplied by the copier manufacturer. This applies chiefly to copiers of the xerographic type, but in practice you may have trouble with the paper feed if you try to save extra pence by buying cheaper grades of paper from a paper merchant. This is not always the case, but the responsibility is yours.

1 SPIRIT DUPLICATORS The paper used for spirit duplication must be fairly absorbent. Too absorbent a paper will pick up too much spirit and cause the dye to run as well as taking longer to dry out. A non-absorbent paper will hold too much spirit on the surface causing the dye to run. Equipment manufacturers supply paper of the right grades and usually will be pleased to preprint them with your own form designs.

Virtually all spirit duplicating papers can be printed on a small-offset duplicator and users who have an internal print department often print their own. Suitable paper can be bought from a paper merchant and this is more economical than buying from a spirit duplicator manufacturer. It is, however, necessary to tell the paper merchant what you need the paper for.

2 STENCIL DUPLICATING   The range of papers which can be used on a stencil duplicator is fairly limited although some manufacturers claim that their present machines are capable of using a wider range of papers than formerly. This is true, but the paper must still be reasonably absorbent. When duplicating on to banks or bonds, set-off (i.e. transferring some of the ink on to the back of the previous sheet) can be largely prevented by using an anti-set-off powder, sprayed on by an accessory known as a 'puffer'. This can only be done satisfactorily on electric machines as the puff of powder is timed by a gear driven from the main drive shaft.

The following remarks apply to all duplicating processes. The equipment manufacturer likes to supply you with paper. After all you are using it all the time and it gives him a steady source of income, whereas you only buy a machine once in several years. In some ways it is safer to buy from the company which made the machine. It is in their interest to see that it keeps working satisfactorily and one way of doing this is to ensure that the paper you buy is suitable for it.

On the other hand you can often buy paper more cheaply from a paper merchant especially if you use a fair quantity over a period; say, at least a ton a year. Provided your order is large enough to be worth having, a paper merchant will look after you very well and he will be able to offer a wider range of papers than the average machine manufacturer. Particularly if you have an internal print department it may be worth your buying other papers as well in this way.

For an experienced operator the chief limitation, apart from size, is that of weight. Very light papers give trouble in feeding, as do heavy cards. The range of weights of paper which can be fed will be given in the manufacturer's handbook.

PAPER, WEIGHT   Paper is sold by weight, and the amount which you get for your money depends on the substance (i.e. the weight of an individual sheet of a given size). The old practice is to indicate substance by the weight of a stock size ream, a ream usually being 500 sheets, but even this can vary. There are a number of sizes known by charming names such as Large Post, Double Cap, Imperial, Royal, Demy, etc. The size we know in offices as Quarto is, for example, a quarter of a sheet of Large Post. All this involves a considerable amount of calculation in order to compare the cost of using two or more different papers for the same job. To straighten all this out the continental system of calculating paper substance in grams per square metre has now been adopted. This is quite straightforward. If, for example, you need 200 sheets to duplicate a job which is being run on A4 size paper, all you need to know is as follows:

1 The substance of the paper in grams per square metre (g.s.m.).
2 That you get 16 sheets A4 size from a square metre of paper.
3 The price per kilogram of the paper to be used.

All you then have to do is to multiply the number of sheets by the substance in g.s.m. and divide by 16 to obtain the weight of paper required for the job. Comparing the price of various alternative papers on which the job could be run is then a matter of simple arithmetic.

PLATES, OFFSET (See also Master.)   Plates used in small-offset work may be either direct image or made by one of the indirect methods. In office duplicating, as compared to internal print department work, they are frequently the former.
  1 DIRECT IMAGE PLATES These may be made of paper, plastic or metal. The last named are used less frequently today than was the case formerly, being less easy to type, write or draw on. However, modern plastic plates have most of the characteristics of a paper plate, being very similar to 'speciality materials' used in the drawing office, i.e. drawing film. Although they cost rather more than the best grade of paper plates they are capable of producing considerably more copies. There is also another type of 'plastic' plate which is paper backed.
      In writing or drawing, professional artists usually prefer to use ink in solid form but most other people find the liquid ready-mixed kind preferable. Lithographic ink which has to

have a greasy content can be obtained in either form and in addition in ball point pens. The final appearance of copies made from a plate on which lithographic ink has been used will depend more on the grease content of the ink than on the appearance of the drawing as seen on the plate, so do not worry if it does not look black enough. The purpose of the ink you are putting on is to pick up duplicator ink. Small-offset being a planographic process, you do not want to indent the surface of the plate as the indentation may not fill with ink on the machine, giving hollow letters.

As an alternative, lithographic pencils are available and non-reproducing pencils may be used for making guidelines.

Direct image plates are sensitive to moisture and should only be handled at the edges. Moisture marks made by the hands are difficult to get out and it is advisable to use a protective sheet.

Corrections should be made in accordance with the instructions given by the manufacturer of the plates. For some a special rubber is used and for others a glass brush, depending on the surface. Using the wrong kind of correction will often spoil the plate. Except in the case of short run plates the plate should be fixed as soon after completion as possible. The appropriate proprietary fixing solution should be used. A thin even film of fixing solution should be deposited by saturating a sponge in the solution and then squeezing it out until it is damp. This is then worked gently over the entire surface. When this fixing process is not required it will say so on the instruction sheet sent with the plates. However, there are some systems duplicators which are equipped with automatic fixing: when using these follow the manufacturer's instructions.

2 INDIRECT IMAGING OF PLATES (See separate entry.)

POSITIVE Strictly speaking, a positive copy is one on which both the tone and direction is right reading, i.e. a black on white copy of a black on white original which reads left to right (assuming the text is in English or another European language). In practice the term can be taken to mean a final usable copy of the original.

POWDER Variously called 'developer powder' and 'dusting powder' the word 'powder' usually refers to the powdered black substance used in electrostatic copiers to make the image visible. Whatever powder is used it must have two properties:

1 Be capable of being charged electrically so that it is attracted only to the image area.
2 Be capable of being fused under moderate heat so that it sets to give a black or near black image, ideally giving the final copy a similar appearance to a printed one.

REAM   Generally a ream is understood to be a package containing 500 sheets of paper. For some unknown reason stationers sell paper by the ream and not by weight, but you still need to know the size of sheet in the ream. For business purposes this is normally A4 or A5 now that we are using International Paper Sizes. You also need to know the substance of the paper which is expressed in grams per square metre. For the same paper you will pay more for a ream for each increase in substance.

ROLLERS (See Cylinders.)

ROTARY   This refers to the type of copier where the original to be copied is passed over a system of rollers which convey it past the light source. In some cases one of the rollers is itself a translucent cover to the light source. With this type of machine the original is taken into the machine and ejected after exposure. Documents of any length can be copied but their width is limited by the width of the feed mechanism.

It is sometimes said, with a good deal of truth, that a flatbed machine requires an experienced operator to get the best out of it while rotary machines are best when they are used by members of the general office staff.

Nearly all duplicators, other than copier/duplicators, are of the rotary type. Regardless of the process used the original is attached to a cylinder which rotates either directly in contact with the copy paper or in contact with another cylinder to which the image is transferred, known as offset duplicating.

SMALL OFFSET   This depends on the fact that water and grease repel each other. The grease is in the ink on the plate (or master) and the plate is damped to prevent the ink spreading out of the image area. On machines designed for use in the office most of the skill in holding a balance between inking and damping has been taken out by the machine design.

Nevertheless this is a simplified printing process and needs more care than other duplicating processes. With a careful and experienced operator small-offset is cheaper, more versatile and produces better results than any other duplicating process. Carelessness produces a wasteful mess.

One advantage is that plates can be made either direct (by typing, writing or drawing on the surface using lithographic materials) or by a number of indirect methods. The quality and hence price of plates varies from paper suitable for a few copies up to the virtually indestructible. Methods of making plates include electrostatic and several 'short-cut' chemical methods which produce inexpensive plates in seconds rather than minutes.

For copies a wide variety of weights and qualities of paper can be used and other materials also when the operator is skilled enough.

Equipment ranges from simple machines for use in the office capable of giving good results on straightforward jobs up to A4 plus size to sophisticated machines capable of full colour process work on large format paper; printing machines in all but name. But the latter do need experienced operators.

STENCIL   This is the workhorse of British office duplicating. The process relies on an ink which dries by absorption into the copy paper. The stencil (master) has a coating which is impervious to ink and a backing which lets ink through. Cutting a stencil is really cutting away the coating in the image area. The ink is squeezed through the stencil from the back.

To enable complex originals to be copied onto stencils the electronic stencil cutter was introduced. The machine consists of two cylinders mounted end to end. To one of these the original is attached. A photoelectric cell scans the original which is attached to one cylinder. This activates a spark gap whenever the cell senses a dark area. In this way line drawings and even halftones can be copied. The equipment is expensive but stencil cutting is available as a service. Stencils which can be cut on a heat transfer copier are a more recent development. They cost about twice as much as normal stencils.

Economical where at least 50 copies are needed, stencil is more expensive to operate than small-offset, less versatile and more restricted. On the other hand it is easier to use and the equipment is considerably less expensive.

SPIRIT   The principle is that a spirit soluble dye is transferred on to the master from a hectographic carbon. This is then wrapped round the master cylinder on the duplicator and the copy paper moistened with the solvent. Hence some dye transfers to the copy paper as it passes in contact with the master. Carbons are available in three grades to make about 35, 100 or 300 copies, these numbers being very approximate.

Good office discipline and the use of carbon sets or edged carbons will prevent the 'purple mess' and you can obtain carbons in black and other colours as well as in purple.

Useful for short-run duplicating where cost is the primary consideration and for systems work. The process is not fully appreciated in this country while abroad it is used for a variety of jobs including duplicating computer output.

TYPING   The method of typing masters varies according to the kind of master you are typing.

1 SMALL-OFFSET   When typing an offset direct image plate it is necessary to use a lithographic ribbon of which four grades are available:

a Fabric ribbons which are suitable for general reprographic work and fine fabric ribbons which should be used on typewriters having the smaller typefaces.

b Pure silk ribbons used for high quality work.

c Carbon paper ribbons which give a fine sharp impression.

d Plastic backed ribbons which give the sharpest impression of all.

The first two types may be used a number of times but the latter can be used once only. Further, these last two should be used only on typewriters which are equipped to accommodate them and are unsuitable when typing on some kinds of short-run plates. The maker's instructions should, therefore, be consulted before using one time ribbons.

Always use a light touch when preparing plates. Heavy pressure will cause indentations which will reproduce as hollow letters. It goes without saying that an even touch is necessary to produce the best results. Electric typewriters give the best results but only if they are kept in proper adjustment, so electric machines used for platemaking should be serviced as regularly as manual machines.

2 STENCIL   When typing on a stencil master you need to break the surface of the wax coating without breaking the porous backing sheet on which the coating has been spread by the manufacturer. Stencils are cut with the ribbon inoperative and the master is usually provided in the form of a set comprising the master itself and a separate sheet with an interleaved carbon. After completing, the backing sheet is removed before the stencil is placed on the machine.

Corrections are made by replacing the surface coating in the affected area by a special correcting fluid. This is spread on the surface with a special brush provided with the fluid. Ideally this should be the same thickness as the coating which it replaces but this is difficult if not impossible in practice. However, piling several thickness of correcting fluid on one area will only make matters worse, the correction probably not coming out properly due to the fact that the corrected surface has not been broken when the correct symbols are overtyped on it.

3 SPIRIT   When typing on a spirit master you are transferring a dye from the hectographic carbon to the back of the master, where anything you type appears as a reverse image. The right reading image transferred by the normal typewriter ribbon on to the front of the sheet has no function in the duplicating process and is only there as a visual check. Provided the touch is heavy enough to transfer the coating from the carbon to the back of the sheet and is even, this is all that is necessary.

4 COLD TYPE   This properly speaking means type set on a typewriter or one of the machines derived from a typewriter. These machines, operated by a typewriter keyboard, are designed to permit variations in the typed work and usually have the type on a changeable segment (Varityper) or on a single changeable spherical head (IBM). Some of these machines enable the text to be proportionally spaced (i.e. spaced as in print) and on some variations in type size are possible as well as type style, by changing the segments or spherical head. Right-hand margin justification is also a possibility although this normally requires a second typing.

Another kind of cold type composer is similar in function to a copier. A fount, i.e. all the letters, figures, etc. in the same size and style, is supplied with the machine. All these are then composed into whatever message is needed, usually a headline or title, and positive or negative copies made. A similar type of machine works on the principle of a photographic enlarger. With this only one size of lettering is necessary and this can be positioned and enlarged to fit a required space.

Photocomposing and 'hot metal' composing machines are expensive and complex and are, therefore, outside the scope of a duplicating unit.

Type composed in any of the ways mentioned may be used for stencil duplicating by using an electronic stencil cutter after composing the type.

CHAPTER 2

Copying
and the management structure

The traditional management structure is like a family tree with the owner as the root and a straight trunk branching out into sales and production. When a company is small this works well enough. The owner is able to supervise the day-to-day operation of the company and little is needed in the way of visual communication. Unfortunately a company of this size, with perhaps up to ten employees, with few exceptions, has to grow or wither. As it grows it develops into something more like a bush than a tree. The original owner becomes the chairman or managing director, and being unable any longer to supervise the day-to-day running of all the various departments, he appoints a board of directors who become the root system of the organization. It is they who decide the policy and who make decisions on all exceptions, but already we have two tiers in the decision making body. Usually the managing director reserves the right to make the final decision on anything which will affect a considerable amount of the company's assets. By now each of these directors has a department for which he is responsible, and as the size of the organization increases each department has to be divided into a number of units, each responsible for one section of the work. (See Figure 1.)

If the original structure is adhered to strictly and the directors delegate little or no authority to those below them, lines of communication become very long and the company structure becomes very rigid. Any decision involving an exception to normal company working has to go up through one department to the director in charge and down through the department headed by one of his colleagues, then back through the same channel until eventually it gets back to the person who first asked for the decision.

As an example, suppose that a salesman working for a company with this kind of management structure met a customer who wanted a product finished in a non-standard colour. His enquiry

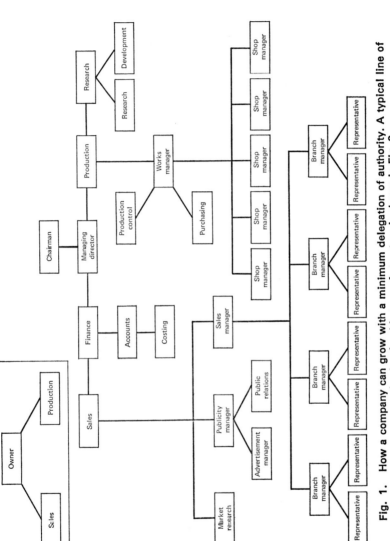

Fig. 1. How a company can grow with a minimum delegation of authority. A typical line of communication for this type of structure is shown in Fig. 2.

whether this could be done would go first to his branch manager, from him to the sales manager, then to the sales director, possibly from him to the managing director. He would pass it to the production director, who would pass it to the works manager, who in turn would pass it to the finishing shop foreman. The finishing shop foreman would then state what was involved and pass the information back through the same chain to the production director, who would then pass the information to the sales director to decide what price increase if any should be quoted. In the case of a large quantity the managing director may also be consulted and it is quite possible that the financial director may also be brought in. When a decision had been taken this would be passed down again through the sales and branch managers to the representative concerned. In the meantime the customer has probably lost interest and bought elsewhere.

If you look at this from a communications point of view the chain of communication is in the form of a large 'V' inverted. (See Figure 2.) Not very many copies are needed because those through whom the query passes are expected to act in sequence and not consecutively. The only real need for copies is among the board of

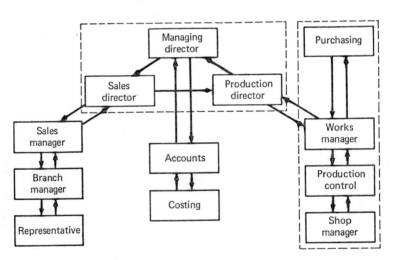

Fig. 2.   A line of communication for a representative's query involving a possible minor modification to scheduled production. The areas enclosed by broken line boxes are those where some speeding up of decision taking can be expected by providing copies of the original copy so that simultaneous action can be taken or consideration given.

directors where two, three or possibly four are involved in making the decision.

However, it would be possible to come to a decision far more quickly if the sales manager, or even the branch manager, had been authorized to collect all the facts needed to make that decision. Further, the time factor would have been cut very considerably if he had been in a position to collect them simultaneously. This, of course, involves cutting across the management tree at about the middle level, or slightly below. (See Figure 3.) The lines of communication are now horizontal rather than vertical. He would get information on production scheduling from the works manager, on costing from the costing department and on other factors such as special delivery, etc. from the appropriate section head. Having collected all the relevant information he would then make the decision whether he could put the order through on his own responsibility. Even if it were necessary to go to higher management the query would go forward with all the facts readily assembled so that an immediate decision could be taken. Depending on the number of departments involved this would probably require a few more copies of the original enquiry but these are a small price to pay for the vastly faster decision.

This is an example of management by exception. (See Figure 3.) In short, higher management only becomes involved in the day-to-day running of the business when there are exceptional circumstances. Now let us suppose that the order which our original salesman received was for items from the standard product range. If they had to go up the sales management ladder and down the production management ladder before they could be executed the procedure would be so clumsy that the business would quickly grind to a halt. Yet this is the way things probably started with the original owner insisting on seeing every order. The probability is that in an averaged sized firm they will be routed via the sales manager although they will be dealt with in the sales office by his assistants. In larger organizations the branch manager will be responsible for this function. It will be his responsibility to see that all the appropriate departments have sufficient information to fulfil the order and to obtain payment for it. To do this efficiently is the object of all order documentation systems. In the past these have involved a considerable amount of transcription but it is now possible to copy a very large part of the information required from

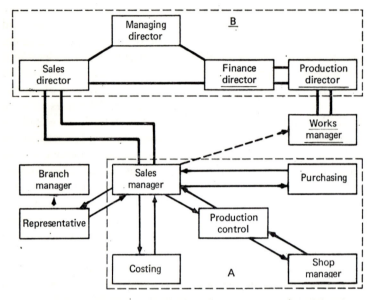

A The sales manager collects information from all departments concerned.

B On receiving this information he decides if in the light of his terms of reference he can accept the responsibility for servicing the enquiry if it becomes an order.

C Only if he cannot does he refer it to the board 'Blue lines'.

D The branch manager and works manager who are not at this stage expected to be directly involved are sent copies for information. They would act on these if they have facts unknown to the sales manager.

The areas in which copies for all concerned are necessary are again defined by dotted boxes. Both could well be serviced in this case by the sales office. Note however that whereas in area A action will be taken on receipt of a copy of the representative's original enquiry in area B a copy of the enquiry plus all the information originating in area A will be needed.

Fig. 3. Possible method of handling a similar enquiry where more authority has been delegated.

one original. This again involves horizontal lines of communication which operate simultaneously between a number of departments. Other similar lines of communication control routine purchasing, stock and production control. At some point these lines of communication interlock; since it is not possible to carry out any of these functions in a vacuum they are all interdependent. Further, the information generated in each is of vital importance to higher management although in a business of any size its bulk is such that it needs to be presented in digest form in order to be comprehensible.

To employ copying to a maximum advantage it is necessary to identify these lines of communication and then to examine them to see how many links in the chain can be served by the copying function. At this stage the form the copying takes is not important; it may be photocopying, duplicating or microfilm.

However, you will find that in every line of communication there are stages where calculation is necessary and others, usually the majority, where all that is required is to convey all or part of the information on a document to others who are required to act on it. In some cases there is a need to combine all or a part of the information from two or more documents. All this can be achieved by using modern copying systems.

To achieve this requires a completely different attitude towards copying than the traditional one of viewing copying as a service to be used or not at the whim of the staff concerned. This was probably fair enough when copying machines were crude and unreliable, when if you needed a copy urgently it was nearly as quick to get a copy typist to 'bash one out' as it was to use a photocopier—but we have progressed since those times.

More recently it was fashionable to talk about the merits and demerits of centralized and decentralized copying services and this attitude still persists in many managements. This is really begging the question. The important thing is how the copying fits into the overall documentation system and this must govern the location of the copying facilities provided. If you examine any line of communication you will find that there are some stages at which it is necessary for the efficient overall running of the system to provide information to those who are to act on it as soon as it is available. In other situations a time lag of an hour or two is not important. This question of timing will not affect every industry in the same way. For example, if you are dealing in perishable goods minutes count in informing the works what to despatch to each customer. In the case of some engineering industries a delay of an hour or two may be quite acceptable. Much depends on individual circumstances. The only generalization which can be made is that where no delay can be tolerated the copying facility needs to be adjacent to the position where the information becomes available. Where this is not so important the copying facility may be at a distance.

The other important factor which results from integrating copying with the overall paperwork system is that each copy made has a purpose. The practice of making copies 'just in case they may be needed' is discouraged. It is the making of such copies under a general purpose copying service plan that has been responsible for much of the increased copying costs about which complaints are heard from time to time. Such copies serve no purpose other than to put money into the hands of the copying supplies manufacturers and to clutter up the files, in many organizations already bulging with papers which have been kept without purpose.

Further, such general purpose copying services are difficult to supervise and are consequently open to abuse. This is particularly true of modern electrophotographic equipment which is so easy to operate that anyone can get a good copy without training or practice. Where machines of this type are open to use by anyone the temptation to make copies for various non-business purposes is very considerable among junior members of the staff. You can easily find that you are subsidising a local amateur football club, the latest pop idol fan club and any other leisure activity your staff happen to be interested in, to the extent of providing them with all their not inconsiderable copying requirements free. You do not usually let junior members of the staff keep the accounts of any clubs to which they happen to belong on the company's accounting machinery, so why provide them with a free copying and duplicating service?

However, this is not the main argument for integrating copying into the overall paper handling routine. Either a given job can be done most efficiently by employing a copying technique or it is done more efficiently in another way. You do not usually give the invoicing clerk the option of preparing invoices by hand or using an invoicing computer. You installed the invoicing computer because it did the job more efficiently, otherwise it would not be there. In the same way if you have a copying facility it should be there for a purpose. It should carry out one or more stages in the documentation routine more efficiently than would be possible without it and staff should use it as a matter of routine.

This, then, is the management background to copying. Admittedly there is much more to management philosophy than this but this is not the place to get involved in a subject on which several separate books could be written. Several points emerge:

1 The copying need is not only governed by the size of the organization but the degree of delegation plays a large part as well. The more management is decentralized and the greater the responsibility delegated to middle management, the greater the need for copying facilities.

2 Where this type of management is practised, lines of communication which the copying services serve are in a different direction to those in a highly centralized business where all decisions, even minor ones, come from the top.

3 The copying facilities provided should be integrated with the overall paperwork system.

4 Copying facilities, and, in copying, duplicating, microfilm and small-offset printing are included, should be treated in the same way as any other management or office tool. There is no basic difference between a piece of office equipment which calculates and one which copies so far as its use in the running of a business or other administration is concerned. One processes data while the other processes words.

5 Once copying is accepted as a basic administrative function in the same way as calculation, and only then, can efficient use be made of the copying function.

In copying, therefore, we have an efficient administrative aid. The actual tools by which it is carried out vary widely in the same way that the tools which we use in the office for calculating vary from a simple adding machine to a computer. In fact, there is a very close analogy between data processing and word processing. If you think of a central reprographic unit as the central processor and backing store then the individual copying facilities are the peripherals. Some of these may be self-contained producing copies and relying on the central unit only for supplies, while others may

be means of access to the central unit that is designed to produce a
master or original from which copies are made in the central reprographic unit. Additionally, there may be facilities which provide a
direct link between data processing and copying. Punched tape or
edge-punched cards may be used to transmit information which,
when printed out, becomes an original which is copied for the next
step in the communication chain. In some circumstances optical
character recognition equipment can be used so that copies can be
turned into machine language for data processing. Telecopying,
that is the transmission of information electronically and the automatic copying of that information at either end of the telecommunication link, is another possibility which has recently become
practical.

With all these alternatives to choose from it is small wonder that
some organizations take the view that to provide an overall copying service is the easiest way out. However, this is seldom the most
economical and frequently does not provide the fastest method of
getting the information to those who have to act on it. To do this
it is necessary to treat copying as an integral part of the overall
visual communication system.

CHAPTER 3

Counting the cost

Just as it is customary still to talk about copying services so one frequently hears talk about the cost per copy as if this were the be-all and end-all of copying costs. Admittedly, the cost per copy can vary widely from one process to another and even within one process when comparing the cost of copies made on machines supplied by one manufacturer to those of another. However, this cost per copy is only a part of the overall copying cost.

Copying, as I have already pointed out, is a part of the visual communication system and the true cost of making a copy must include the 'wasted' time between when the original is available for copying and the time when the copy reaches the person who has to act on it. The word 'wasted' in this context means any time which it would have been beneficial for the organization to have saved. If, for example, it is immaterial whether the copy arrived at its destination within 2 minutes or 2 hours of the original becoming available, then any time within 2 hours has not been wasted. There is, however, one other kind of wasted time, that is the time taken by an employee to make the copy when he or she may have been doing something more useful.

It is one or other of these elements of wasted time, or the two together, which makes copying expensive. This in turn makes the siting of copying facilities particularly important and is the true argument in favour of siting them adjacent to the point where they are needed, even if this means that they will sometimes be under-utilized. It is only once this has been accepted that the cost per copy becomes a factor which is relevant to the overall cost. (See Figure 4.)

This can best be illustrated by a slightly exaggerated example. Suppose that you have to get out 10,000 statements every month. This can be done in a number of ways, one of which is to copy them on a xerographic machine using the customer's account with an overlay. Using the fastest copier/duplicator you can accomplish this task in under 3 hours. The same task could easily take a month

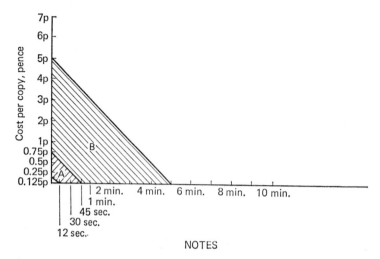

NOTES

1   Area A represents the time spread taken by various types of copiers to make a single copy, excluding copier duplicators. This varies from about 12 sec for an average electrostatic to 45 sec for a slow diffusion transfer copier.

2   Area B represents the added time which may be taken between the time the person making the copy picks up the original and resumes other work.

3   The diagram is intended to show that convenient placing of copiers is important and that where they are at a distance from the work station of the person requiring the copies the time taken going to and from the copier can be disproportionate.

Fig. 4.   The labour content of copy costs, the labour cost unit being 1 penny per minute.

using normal transcription, i.e. typing the information on to statements and then checking it. Now on average all your statements would be 2 weeks late going out. Assuming that your customers pay on time that would mean that you had to borrow for 2 weeks the value of the work represented by your invoices. Suppose the average value of an invoice is £100. Even if you can borrow this money at 7 per cent this would cost you £2916.75 every fortnight during which your statements were overdue. Granted that in practice the difference would not be so great, you are still losing a disproportionate amount of money when compared to the cost per copy of duplicating, even if this entails paying a maximum of 1½p per copy. You can if you prefer look at it in a different way. In order to get your statements out on the first day you would need to take a certain number of copy typists away from other work. The directly measurable cost of doing this is the cost per hour of employing a copy typist multiplied by the number required to do the job, multiplied by the number of hours they are employed on the task, plus the cost of checking their work worked out on the same basis. In addition there is a hidden cost in delaying the work they would normally be doing. This kind of calculation you can easily work out for yourself but if it comes to less than 12½p per statement you will be lucky.

In the example I deliberately chose an exaggerated situation to show that transcribing from one piece of paper to another, which is cheap in material costs, is one of the most expensive ways of copying when the overall cost is counted.

The materials cost per copy of A4 size is the usual standard used when comparing the cost of one process with another and for the processes in more general use these costs are given in Figure 5. However, even if the overall cost in the terms already discussed is ignored, these still do not give a true cost per copy. To do this it is necessary to take into consideration both the cost of the equipment used, the cost of the space which it occupies and that of wasted materials.

All these vary widely with different processes and in different situations. Always bearing in mind that a high cost per copy may be acceptable because it makes a larger saving elsewhere, the part played by overhead costs is largely a matter of utilization. Some years ago when all copiers used one or another of the chemical processes equipment costs were low; it was an expensive machine which cost £200 and many cost less than £100. In these circumstances even if you write off the machine in three years it did not need many copies to be made each working day to reduce the overheads attributable to each copy to negligible proportions. (See Figure 6.)

Today, when so many electrophotographic machines are in use, the position is a little different. This type of machine is by its nature expensive to make as precision engineering throughout is necessary if it is to function satisfactorlly. Consequently prices for this kind of machine start at about £175 to £200 and many machines cost somewhere nearer £500. Let us say, for example, that you only need to make 200 copies per week. A diffusion transfer machine which will do the work you require costs, say, £75, an electro-

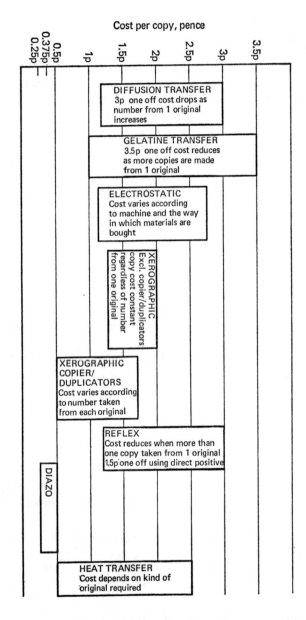

Cost per copy, pence

0.25p
0.375p
0.5p
1p
1.5p
2p
2.5p
3p
3.5p

DIFFUSION TRANSFER
3p one off cost drops as number from 1 original increases

GELATINE TRANSFER
3.5p one off cost reduces as more copies are made from 1 original

ELECTROSTATIC
Cost varies according to machine and the way in which materials are bought

XEROGRAPHIC
Excl. copier/duplicators copy cost constant regardless of number from one original

XEROGRAPHIC COPIER/ DUPLICATORS
Cost varies according to number taken from each original

REFLEX
Cost reduces when more than one copy taken from 1 original 1.5p one off using direct positive

DIAZO

HEAT TRANSFER
Cost depends on kind of original required

Fig. 5.   The approximate materials only cost per A4 sized copy.

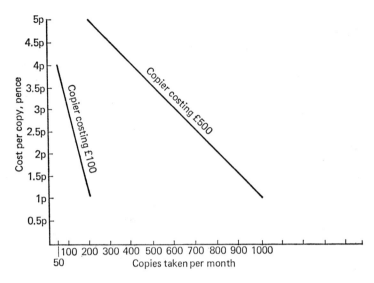

Fig. 6. The cost per copy represented by capital outlay on a
copier written-off over a five-year period.

photographic one, say, £375. Suppose, for argument's sake, that you
decide to write off the machine in 3 years, whichever it was you
chose. In the case of the diffusion transfer machine this would be
£75 divided by 31,200 or 0·24p per copy. However, if you decide
to use the electrophotographic machine the cost will be 1·21p, an
appreciable addition to materials cost per copy.

This example is given to show that the cost of the equipment can
be important whatever the other attractions of a particular process.
If we take this exercise one stage further we will see that the
materials cost per copy for diffusion transfer is 2·5p for the first off
and 1·10p for subsequent copies, whereas for electrophotographic
it is 1·5p per copy. If, therefore, all the copies to be made were first-
off copies, in the example given, the overall cost per copy using
diffusion transfer would be 2·74p compared to 2·46p using the
electrophotographic machine. In other words the overall cost per
copy would be slightly lower in spite of the machine itself costing
five times as much. If, however, the copies were to be made in the
ratio of one-fifth first-off copies and four-fifths secondary copies the
position would be reversed and diffusion transfer would be cheaper,
the average overall cost using this method being 1·65p compared

to 2·46p for the electrophotographic machine, or nearly 1p per copy less.

In this example the figures have been chosen to illustrate the three points which come out. Other machines using either process would give a different result, particularly where one of the most expensive diffusion transfer machines is compared to one of the cheapest electrophotographic ones.

However, not all copying processes give good or even readable copies every time. In fact it is the exception rather than the rule to find a copier which is infallible in spite of some of the claims which have been made by overenthusiastic salesmen. The reliability of any particular process has therefore to be taken into account when costs are being considered. Basically electrophotographic processes are more reliable than the older chemical ones and, where electrophotographic equipment is concerned, it is more often the design faults in individual pieces of equipment which give trouble rather than any fault in the process itself. Although this may be cold comfort for those who have suffered from equipment which gives trouble it nevertheless argues well for the future when we can look forward to trouble-free equipment. It is often overlooked that the electrophotographic copying revolution, if such a strong term can be used about a copying process, has taken place in less than a decade, mostly since the mid 1960s. During this period a very large number of different models have appeared, some hastily designed by companies who have no previous experience of either electronics or copying. It is small wonder that some have not lived up to the claims made for them. These remarks do not apply to the majority of electrophotographic equipment and when the market settles down again machines which are mechanically unreliable will undoubtedly no longer be with us.

Now that we are coming into the area of applications copying, where the copying forms a part of a system, unreliability is becoming potentially even more expensive. It is no longer a mere question of wasted materials but of holding up a complete paperwork system while faulty copies are remade or, in the case of an equipment breakdown, while this is being put right. This brings us to another aspect of cost which is purely concerned with servicing. When copying, or for that matter other equipment, is being used as a part of a paperwork system it needs to be as reliable as any machinery used on a production line; reliable not only in the sense that it continues to function but that it reliably produces the result for which it was installed. Otherwise the results could be very serious, affecting procedures which are themselves remote from the copying itself. For example, a breakdown in a copier used to produce monthly statements could result in your statements going out late, with the consequence that you have to borrow money to cover the period between the time when you would have received the money if your statements went out on time and the time when you actually got the money in.

This kind of expense would be difficult to prove, since there is always an element of doubt but it is one which has always to be borne in mind. It is not an argument against using copying as part of an overall paperwork system since the money saved thereby can outweigh all other considerations. It is, however, a very good argument for ensuring that prompt and reliable service is always available for the equipment which you buy and wherever possible for covering yourself by ensuring that alternative equipment can be used in an emergency. Admittedly many smaller firms cannot afford to have surplus pieces of copying equipment on their premises and even large ones would find the duplication of equipment prohibitively expensive. But there are always copying services available these days and if one of these is operated by the manufacturer of the equipment you purchase, it is not unreasonable to expect some consideration in using it should your own equipment break down at an inconvenient time.

Once you accept copying as a part of an overall paperwork system it is necessary to treat the equipment as a management tool in the same way as an accounting machine and to consider the merits and demerits of the various machines offered in the same way, reliability, service and fitness for purpose being more important than gimmicks.

We have, therefore, four aspects by which costs can be judged:

1 There is the materials cost per copy which, although important, tells only a part of the story.
2 There is the overall cost per copy which in addition to materials, takes into consideration the cost of the machine itself, the space it takes up and the cost of operating it, i.e. every cost which occurs between the time someone drops what they are doing, goes to the machine, makes a copy and

D

returns to take up what they were doing previously. It is this cost which is important when deciding where to place general purpose copiers.

3 There is the systems cost of copying, that is to say the cost of carrying out a step in a paperwork system using a copying method compared to other methods available. It is this cost which is of paramount importance when the copier is being used for one particular application.

4 There is the by-product cost. This comes into consideration where a copier is to be used to make masters for a secondary copying or duplicating process as well as for straightforward copying. Generally these are diazo, spirit or small-offset duplication masters. In these cases the cost of making the master may be higher than when using an alternative process, but the cost of making the master plus the copies made by the secondary process may well be less than when making all the copies by an alternative copying process. However, when this is being considered it should be borne in mind that some originals may not be suitable for making masters by the secondary process and all copies from these will need to be made by the process normally used to make the masters.

Material costs per copy

It is these materials costs per copy which are usually quoted in reference works largely because they are easy to work out with a fair degree of accuracy and are valid in all situations. Other aspects of costing listed above are more difficult to work out and are valid only in individual circumstances. Nevertheless the amount of time spent by staff in going to and from the nearest copier or the money saved by incorporating a copying step in an office system may be far more important than the actual cost per copy. Therefore, although it is necessary to consider carefully the materials cost per copy in conjunction with other characteristics of each process capable of doing a particular copying job, this is not the sole criterion.

CHAPTER 4

Systems or application copying

The serious application of systems copying is comparatively new. For many years ingenious office managers have been applying copying to a special function and making their own masks and overlays but it is only since the early 1960s that a manufacturer started to sell equipment on the basis that it would be suitable for a particular paperwork function.

The first equipment to be sold in this way was photostabilization, or reflex. This was at a time when electrostatic copying was still a rather expensive novelty and reflex had lost ground to diffusion transfer. In fact salesmen for diffusion transfer machines would tell you confidentially that reflex was not to be considered seriously as an office copying process.

About that period at least one manufacturer of reflex equipment held demonstrations to show that the ability of a photostabilization negative to be stored and reused at a later date could be put to good use in a number of accounting applications. One such demonstration involved taking copies of an invoice and then cutting the pricing column off to produce a delivery note. If memory serves well a strip of photographic negative was used on which the words 'delivery note' were printed in bold type. Another part of the demonstration involved opaqueing out some of the image on the negative and in another, two negatives were copied together. All this was very ingenious and, if electrostatic copiers had not reached their present state of development, it is possible that reflex copying would have been developed further along these lines.

However, xerography is today undoubtedly the leader in the systems copying field. For this it has two advantages. It copies on to plain paper and good copies can be made from copies. In fact a copy from a copy can sometimes be better than a poor original.

Undoubtedly the effort which Rank Xerox have put into promoting the idea of systems copying has had something to do with the adoption of the xerographic process for special applications, but it must be admitted that this would have come to little if the process had not been suitable.

The whole system relies on the use of overlays, sometimes called templates or masks. By changing the overlay it is possible to change one document into another rather in the same way that children sometimes have model figures which can be dressed in the uniforms of different regiments or dolls which can change their nationality. Even nearer to the overlay principle is the identikit which comes into every television crime serial. In this a number of different characteristics of the upper, middle and lower parts of the human face are drawn on separate sections of the pages which are then bound into a loose leaf volume. By turning the separate sections it is possible to build up a face with any desired characteristics, to add or remove beards, moustaches or spectacles. Once the desired face has been selected the complete page is copied to produce a portrait likeness.

In the same way, by using overlays, it is possible to change the characteristics of a document so that it serves another purpose. The overlay itself can be either opaque or transparent, or can have some areas which are opaque on a transparent sheet. It may be either plain or have any desired wording or design printed, typed or written on it. The original may be either copied in full or any part of it can be combined with an overlay or, if necessary, one or more originals can be copied at the same time, as an original. The possibilities are endless.

All this is worthwhile because it saves time, labour and the possibility of transcription errors. By using a copying system it is possible to obtain a document set from a single original in at most two minutes without the possibility of mistakes being made due to hand copying from one document to another. The savings which are possible have been proved by a number of companies both in this country and overseas. In some cases up to 50 per cent monetary savings have been claimed by users when compared to previous methods used. Another advantage mentioned by users who have been interviewed by the author is that the same number of staff are able to handle a considerably larger volume of document sets without overworking or having to work overtime. Cleanliness and not having to stock a variety of preprinted forms are other advantages which have been mentioned by users of xerographic systems.

To be fair, most of the systems which can be carried out by

copying can also be carried out by using a systems duplicator and the pros and cons of the alternative methods are discussed in a later chapter on systems duplicating. Before considering these I will discuss some of the possibilities of adapting xerographic copiers to the production of documentation sets. At the time of writing the overlays have to be moved by hand, but I understand that development work is now going on which will eventually result in the automatic placing of overlays on the platen, thus eliminating a further possibility for error.

Invoicing

The sales documentation set, whether for the home market or export, is costly to produce and liable to error particularly if one of the modern electronic accounting machines is not used. Transcribing information from one document to the next is monotonous work and consequently difficult to perform quickly and accurately.

The conventional method compared with the xerographic method is shown in Figure 7. From this it will be seen that at least four separate typings, all of which need to be checked, are reduced to one typing of a master document. Once this has been checked the information on it is valid for all other subsequent documents. In the procedure illustrated four separate overlays are used for acknowledgement, order assembly, despatch and receipt notes and the invoice itself, but the method is flexible and can be extended to include shipping documents if necessary. Further, by taking the copying function back one step much of the possible area of error in making up the master can be eliminated. For example, names and addresses of regular customers can be copied from a card index and since this is a dry process, labels can be made on gummed paper at the same time. Descriptions of products ordered can be held in a strip index and transferred to a carrier. This possibility is of comparatively limited application but is mentioned here to show that if accurate description is of particular importance it can be done. Even the copying of the actual goods shipped has found applications. Even if an accounting machine is used to produce the master document considerable savings can be made by using copying methods to produce the subsidiary documents.

Most modern electronic accounting machines can be programmed to produce virtually any information desired. Their copying ability,

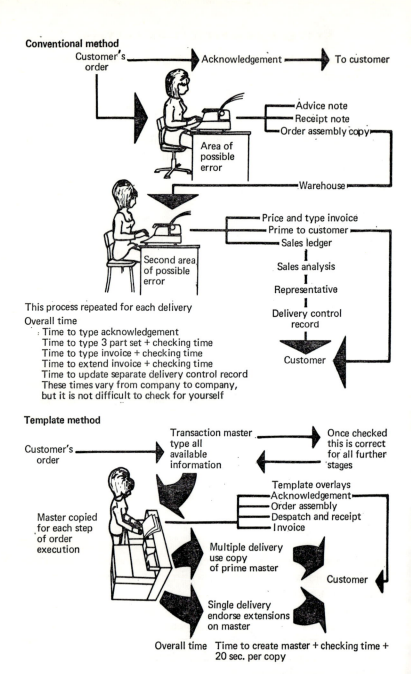

Conventional method

Customer's order → Acknowledgement → To customer

Advice note
Receipt note
Order assembly copy

Area of possible error

Warehouse

Price and type invoice
Prime to customer
Sales ledger

Sales analysis

Representative

Delivery control record

Second area of possible error

Customer

This process repeated for each delivery

Overall time
: Time to type acknowledgement
Time to type 3 part set + checking time
Time to type invoice + checking time
Time to extend invoice + checking time
Time to update separate delivery control record
These times vary from company to company,
but it is not difficult to check for yourself

Template method

Customer's order

Transaction master type all available information

Once checked this is correct for all further stages

Template overlays
Acknowledgement
Order assembly
Despatch and receipt
Invoice

Master copied for each step of order execution

Multiple delivery use copy of prime master

Customer

Single delivery endorse extensions on master

Overall time   Time to create master + checking time +
20 sec. per copy

Fig. 7.   An invoice set prepared from one master not only saves time
but eliminates transcription errors.

on the other hand, is confined to the production of a four-part set. Using modern direct-entry computers it is possible to produce for example, packing information on the invoicing master. An analysis showing gross and net profit on the order is another possibility. From the copying viewpoint it is a question of separating columns of figures into those required on the various documents and adding the appropriate headings. The printing from an electronic invoicing machine or direct-entry computer is usually arranged so that any analysis figures appear in a separate area of the form. The area containing the invoicing information can, therefore, be treated in the usual way to produce an acknowledgement, despatch note (with pricing information masked) and the invoice itself. The analysis can be copied separately either in whole or in part for distribution to those executives who have to take action on it.

A further extension of the use of the invoice copy is in sending out reminders to late payers. In this case it is not necessary to go back to the original master as a copy can be made from a copy and the usual procedure is to take the file copy of the invoice and copy it together with an overlay which is worded so as to turn it into a request for payment. If necessary a series of such overlays can be used, the wording being stronger in tone the later the payment gets.

The same principle has been applied to correspondence. In some kinds of business it is difficult to get a reply to a query sent to customers or suppliers. Instead of sending another letter referring to an earlier one which entails separate typing and checking, it has been found effective to send a copy of the original which has been overlaid with a request in bold type that an immediate reply would be appreciated.

Purchasing is another area in which the copying technique can save time and money. Basically the procedure is similar to invoicing. A master is created which contains all the information necessary for the purchase order. Copies are then taken of this and overlays are used to delete the pricing information on receipt of goods notes, such as are sent to 'goods inwards'.

It is in this field that more opportunities occur for preparing the original master with the aid of copying, since it is the purchaser who is dictating the terms. In this context there are even more opportunities for combining card files of suppliers and commodities. The files can be those created for another purpose other than the copying function, since only the required information needs to be copied from each card, for example, the name and address of the supplier from the record file and the description of a product from a product file. Whatever other information the cards may contain does not matter, even if this is confidential, since it will not copy through an opaque overlay or when the cards have been slotted into one of those slotted listing boards, so that only the required information is presented to the copyboard.

The basic purchasing method is to copy the name and address of the supplier from one index and the products required from a product index. The combined copy then has the quantities and prices filled in on a typewriter and a second copy is then made from the completed copy. This may either be on preprinted paper or the standing information can be added by means of another overlay. Copies are then made for all internal departments needing to know about the purchase and, if necessary, pricing and other information can be deleted on these as well as changing the document title.

There are many variations on this procedure and some industries have found that where the objects purchased are small enough it is helpful to copy these together with the description. Depending on the industry, this can sometimes be used to aid in identifying piece parts or even products distributed to somewhere that samples have not been distributed to. A typical example of this is where goods are despatched straight from a manufacturer to the branches of a retail organization. In the fashion trade, for instance, it is often difficult to give an adequate description by which the goods can be identified with certainty. A copy aids this identification and in some instances it is only necessary to copy a part of the sample to enable the branch positively to identify the model. This is usually some detail which did not appear on the previous season's style. To do this it is, of course, necessary to use a flatbed copier and the fact that this will copy three dimensional objects is often claimed to be a gimmick. It does, however, have uses in some trades and industries.

In some kind of mail order business it is even possible to use the orders sent in by customers as the master from which the acknowledgement and works orders are taken. This may not be possible in every case since some customers will inevitably fill in the order form either indistinctly or incorrectly. However, if say 70 per cent of customers' orders can be treated in this way the savings can be exceptional.

So far the combination of a copier and a printing calculator has not been much exploited but the advantages are considerable. The only snag in using a printing calculator is that it has no narrative ability. It simply prints out a list of figures plus mathematical signs. This means that the printout is only intelligible to those who are familiar with the way in which the calculator is used. It is, however, not difficult to combine all or part of the information on a tally roll with narrative on an overlay. One application for this technique would appear to be in the rapid confirmation of estimates or quotations given over the telephone. Now that the price of electronic calculators is dropping to the level of the electro-mechanical type it is possible to supply senior sales executives with the former type on an individual basis. This, combined with decimalization and metrication, means that a sales manager is able to work out an estimate while talking on the telephone even if the quantities and percentages involved are quite complex. Even with a printing calculator this would require a separately typed letter to confirm because of the absence of narrative, already mentioned. With the aid of a copier the amount of typing involved can be reduced considerably. Where the amount of the quotation is all the information required by the customer this can be copied together with a standard letter in the form of an overlay. It is then only necessary to fill in the customer's name and address plus the date, after which a second copy can be made of the complete quotation, the first with the typed information being retained as the file copy. Extra copies can, of course, be made for representatives, branch managers, etc. and if desirable a further overlay can be used to ensure that their purpose is understood by the recipient. Provided the amount of detail is not too great it should be worthwhile using the copying method even if descriptive matter is required on the quotation.

Invoicing by means of combining tally roll and overlays on a copier is quite feasible. Here the other limitation of a tally roll becomes apparent. The figures will appear in a vertical column, whereas conventionally they would be placed horizontally across the form. However, this is largely a convention and provided the narrative is fully explanatory would certainly not invalidate the document.

Other possible applications for this technique are more specialized but it undoubtedly has applications in several fields including specification writing, stock and production control.

Already copying is being used to transfer engineering specifications into works orders covering some 200 variations in a standard product line, amendments being dealt with by whiting out the affected information and recopying the amended page.

It is not only in industry that application copying has been applied successfully. Insurance companies, which often have difficulty in obtaining replies to their letters, have used the copying technique previously mentioned with considerable success. It would appear too that using the copying technique or a variation of it would considerably cut the time needed for correspondence when making quotations. In particular copying the invitations to quote for motor insurance which appear so frequently in newspapers and magazines and sending the copy of the completed quotation back to the prospective client would be a valid use for the copying technique, the form having been redesigned so that only the figure need be filled in. In addition to saving clerical work this would place the onus of filling the form in correctly fairly and squarely on the shoulders of the prospective client.

Banks too have used copying for a number of purposes. One of the more interesting is that of preparing lists of shares which some banks, stockbrokers and investment companies send regularly to their clients. The list generally remains fairly static, only the prices change. These can be blocked out with a suitable correcting fluid and since only some of the prices change within a given period a list once typed will last for some time, whereas when stencil is used a new master must be typed for each list.

There is also a copying system for the documents used by stockbrokers to complete a deal according to the procedure of the London Stock Exchange, and specially tinted papers are available for this purpose.

The above examples are given to show the wide scope of application copying. Most have not been detailed because the details would vary with every application and slavish copying of a system already used by another organization would not produce the best solution to a problem, and the forms and overlays are best when designed to meet the individual problems of users.

Not all applications involve the use of overlays and it is sometimes sufficient to provide extra copies of a document which then play their part in the paperwork system. In most business applications these are the result of straightforward company administration, but instances do come to light from time to time where a copy is used in place of an original document because the original is a document of title while the copy is not. In other words, while the copy may indicate something about the property in question it does not enable the possessor of the copy to transfer the title from the present owner to someone else. Hence this would appear to have applications in hiring or hire purchase.

Another group of copying possibilities lies in translation. Overlays have already been used by an international electronic group to recode engineering drawings and translate the title so that drawings originating in one country can be used in another. There is no good reason why a similar technique should not be used to translate any document where the narrative is permanent and only the figures change. For that matter there is no reason why an overlay of a despatch note, invoice, etc. should not be in a foreign language with identifying narrative outside the copy area. This would enable overseas documentation to be handled by someone unfamiliar with the language. Although it is often said that one of the difficulties in dealing with overseas customers is that, when they do not speak English, multilingual and expensive translation services have to be used, little attempt appears so far to have been made to utilize this technique to relieve those members of the staff of export departments of routine translations. Or is so much international business done in English these days that it is unnecessary to translate routine documentation? To one not directly engaged in export trade reports on this aspect of overseas business are contradictory.

What can copying undertake? Adapting the idea

All steps in a documentation procedure can be classed as ones in which mathematics are needed and those in which it is merely a question of copying all or some information from one document to another. Generally speaking, the steps which require calculation are the ones which produce primary or master documents, while those requiring the transposition of this information to other documents can be considered secondary, not because they are any less important but simply because they do not contribute any new information.

If we look at the sales documentation procedure again, comparing the overlay method with the conventional one, we will see that on the master we type all the information necessary for the customer's acknowledgement, works order, despatch and receipt notes and the invoice itself. When we receive the order we know how much we are going to receive for it, by what means we are going to despatch it and how it will be packed. There may be one or two details to add, such as the date of despatch, weight of cases, etc., but these originate on a weight note in the warehouse. The only step which involves calculation is the extension of the invoice. The only other document in the set which adds further information to that available when the order is received is the weight note. Therefore, in this set the invoice master and weight note are the two primary documents. All others can be copied from information appearing on one or the other. This applies to the vast majority of goods sold. There may be exceptions, such as products which are hand crafted to special order, but these are comparatively few.

If you go through any documentation procedure, whether it is concerned with buying, production or stock-keeping, you will find a similar situation. On each document you will find static information (usually printed) and information which applies only to that document set. You will also find that in most cases the majority of this variable information originates at one point in time. It then appears either in whole or in part on all the other documents in the set. Applying copying is largely a question of identifying this point in time and then redesigning the document set so that as much of each subsequent document as possible can be completed by copying from the master document.

It is just another one of those things which makes you say, 'That's only common sense,' and, of course, you are right. It is a question of having enough time to get down to studying the problem. In fact some of the leading copying machine manufacturers employ systems analysts to help you. You are entitled to examine any scheme they come up with critically. After all, they are there to win a customer for the salesman they are backing. If they cannot show a saving you are quite entitled to show them the door, nor need you accept the first system presented. Most things which can be done with a copier can be done with a systems duplicator and these will be discussed in the next chapter.

The various systems available

In theory any flatbed copier can be used with overlays to produce the kind of application copying discussed in this chapter, but in practice three processes are much more frequently used than the others—electrostatic, heat (infrared), including dual spectrum, and diazo.

At the time of writing Rank Xerox claim that they are more advanced in applying copying to systems work than their rivals in the copying field. In support of this they point to a multinational exchange of information within their organization whereby the solution to problems originating in one country becomes available to actual and potential customers in all other countries where they operate. Their team of systems analysts is international.

For the xerographic system they claim the following advantages:

1 Copies are on plain paper, which can be preprinted if necessary.
2 Their machines are rented. If you do not like the system or it is not saving you money you can terminate the agreement at short notice.
3 Their machines can be operated by junior staff. Even what they term a 'key operator' can be trained in a day and needs only to know a few simple procedures.
4 They have service mechanics 'everywhere', important when the copying is tied into a paperwork system and a breakdown could cause serious disruption.
5 They have developed a range of copier/duplicators which are particularly suitable for systems work. Some of the smaller of these are only differently metered versions of their standard office copiers but the larger models, complete with automatic collators, approach the speed of operation of mechanical duplicators.

Other electrostatic machines which copy on to zinc oxide coated paper are suitable for systems work, flatbed machines probably being better for this purpose than rotary models.

Most of the leading manufacturers have people on their staff who specialize in systems applications and various accessories are available which are useful in this kind of application. Generally speaking these machines have the same advantages as xerography except that the copy paper is coated. There is a tendency among some manufacturers to recommend a combination of electrostatic copying and systems offset and this will be discussed in the next chapter. At the time of writing no copier/duplicators were available approaching the speed of operation of the xerographic machines.

Dual Spectrum copiers (3M Company) have been used for a number of systems applications. They will copy any original and, as the manufacturers are always emphasizing, this is another 'all dry' process. Coloured copy paper is available, suitable for specific financial applications as well as for colour coding. For situations where quick copies are needed from originals under the control of the person making the copies heat reflex has advantages. A 'blind' colour can be used for any information which is not to appear on the copy, but masking is better carried out on the flatbed machines used for Dual Spectrum.

Diazo still has considerable potential as a systems copying process. I say 'still' because it looked at one time as if it would be *the* systems copying process. Machines are available with automatic stacking for both originals and copies and with a number of other accessories. Where the originals are or can be one sided and reasonably translucent, this is still the least expensive process to use. Quality may not be up to the best electrostatic standards but unless the operator is especially ham-fisted every copy is readable.

CHAPTER 5

Systems duplication

Systems duplication performs roughly the same function as systems copying and for practical purposes they can be considered as alternative methods of achieving the same result.

The two duplicating processes most frequently used are spirit and small-offset. Both have areas in which they are the best solution to a paperwork problem and to some extent these overlap. However, at the time of writing spirit is the only process on which line selection is a practical proposition, so if this is necessary to the system it is a natural choice.

Spirit duplication for systems work

Although it is possible to use general purpose spirit duplicators for simple systems work their limitations for this purpose are obvious. It is, however, the simplicity of the basic machine which has led to the development of a range of special purpose systems machines.

Generally speaking these machines fall into two categories: those which allow random lines on the master to be selected and copied adjacent to each other and those which select certain areas of the master, enabling different parts of it to be copied. In this way a document set can be made from one master just as it can by using overlays on a copier. There are, however, two differences. On spirit machines of this kind the masks which make this possible are incorporated in the machine and operate when the appropriate control is used. Alternatively, on some machines the master is moved on the drum so that only the part containing the required information comes into contact with the copy paper.

As with all spirit systems the 'static' information is added by the use of preprinted stationery. The grades of paper used for spirit duplication print quite satisfactorily on a small-offset press, so you can if you wish print your own forms.

Systems machines naturally cost more than general purpose duplicators, as the selection mechanism must be well engineered if the machine is to give reliable service. How much more depends largely on the sophistication of this selection mechanism, line selection machines being generally the most expensive.

Nearly all systems spirit duplicators sold today are electrically operated and, being intended for continuous operation, are designed so that the user is seated. Some are designed on the principle that, if the selection of certain areas or a number of lines is provided, it is then up to the user in conjunction with the seller to devise the system with which they are to be used. Others are designed with one specific purpose in mind. There is, for example, more than one machine designed to produce a set of shipping documents from one master. These documents are designed to a format agreed with the DTI (then BoT) and accepted at major British ports. Not only do these allow a set of documents to be prepared in minutes, once the master is available, but the possibility of transcription errors is eliminated. The cost per document is lower than when a copying process is used and it is a question of whether you feel that the office discipline necessary to produce acceptable copies and keep the office reasonably clean can be enforced.

This is a special application and the main area for spirit duplication is in stock and production control. As readers who are engaged in industry are aware, this needs a job ticket, stores requisition and route cards, several copies being needed of each, to keep costing, stores and labour records up to date. This applies equally whether the production is one off, small batch or long run.

Because spirit masters can be prepared inexpensively on a typewriter the spirit system is particularly well suited to situations where production is either undertaken in small batches or where variations in the product call for differing selections of piece parts or sub-assemblies. The alternative system based on an addressing machine needs a separate metal plate to be embossed for each part. This is a good system where production is stablized over a long period as once embossed the plate can be reused almost indefinitely.

However, from an engineering point of view, modern production calls for considerable variation over a wide range of products, even a product as standard as a mass produced motor car which is subject to considerable variation during the production run. This is necessary both to give variations in colours and trim and to comply with the requirements of overseas markets. I mention this here as cases have been reported to me where this kind of variation has been overlooked simply because the production control staff were not consulted in time.

Regardless of the product, the basic principle is the same. Instead of writing or typing on plain paper, the production control department types the specification on a spirit master. Line and/or area selection enables this specification to be broken down on to job tickets, stores requisitions and route cards.

Indirect masters

In some cases it is more convenient to type the specification in the normal way and to make the spirit master indirectly. As previously mentioned the most economical way to do this is by using an infrared copier of the transfer type. The number of copies obtained in this way appears to vary considerably. Some companies are using this method in systems work with complete satisfaction. Others have abandoned it after encountering difficulties. Much seems to depend on the quality of the original from which the master is made and on how carefully the copier is used. It is certainly a convenient method where the same specification is repeated at intervals since the typed master can be filed and a separate spirit master made from it for each batch. Normally a spirit master is good for the documentation for one batch only. In the past difficulties have arisen from the printing on the forms being copied. If the variable information is not printed neatly between the print it is partly obliterated by the print on the copy. Unless the system calls for the print on the original form to be copied and to appear on the spirit master, this can be overcome by printing the forms in red or some similar shade which is blind to infrared copiers.

Another use for the copying technique is to enable information which originates at a distance from a duplicator to be incorporated in the spirit duplication system. For example, in a sales office which may be miles away from the factory.

Spirit duplication in sales and accounts

The preparation of a set of export shipping documents from one master has already been mentioned. In a similar way, a set of home market sales documentation can be completed from one master. This set can be complete with acknowledgement to customer,

E

despatch note, and receipt, works order, etc. and with the extension added on a second master, a set of completed invoice forms.

Depending on the accounting machine used it is sometimes possible to prepare the master on the accounting machine itself, rather than to use the indirect method. It is necessary to have the output print head on the accounting machine adapted to take a hectographic ribbon or to use a made-up hectographic carbon set in place of the normal multipart set. In either case the adaptation would need to be done by the machine manufacturer. Some accounting machine manufacturers encourage this but others are not so keen. Whether this is satisfactory depends on the design of the output typewriter on the accounting machine, and any-one considering doing this should first consult the machine manu-facturer.

In theory it should be possible to treat a line printer attached to a computer in the same way but, so far, to the best of my knowledge this has not so far been done. The difficulty is that at the speed at which line printers work it is difficult to get a suffici-ently even impression, although the stringent requirements of OCR (optical character recognition) can be met so the problem is not very great.

Small-offset for systems

Small-offset systems machines like those using the spirit process cost considerably more than general purpose offset duplicators. This is because the machine has been automated to an extent where operating has virtually been reduced to pressing a button. The blanket is cleaned, the old master ejected and a new master mounted, etched and inked, all without the operator having to do anything. This has resulted in a performance in the order of 10 copies from each of 4 masters in 2 minutes. The speed of opera-tion varies from one machine to another, as does the degree of automation, but this can be taken as about average for a sophisti-cated machine.

Readers will note that the speed has been given in terms of so many copies from so many masters in a given period. It is this that counts on systems machines rather than the number of copies which can be made in an hour from the same master.

When considering systems offset machines it should be borne in mind that the concept of their use is quite different from that of the internal print department. Systems offset machines, like any other process used for systems duplicating, is a part of an overall paperwork system and should be treated as such.

If you accept a stencil duplicator or a spirit machine in the office itself there is no reason why a systems offset machine should not be similarly placed. In most, if not all systems applications, it is an advantage to have the machine adjacent to the machine which is producing the masters so that it can complete a paperwork flow smoothly without interruption. This implies that the procedure on which it is used should be important enough and the volume sufficient to justify the cost of a small-offset printing machine. Providing these conditions are met small-offset has a number of advantages over other duplicating systems:

1 Masters (or paper plates) are inexpensive, those made for short runs costing only a penny or two.
2 The quality of copies is good, better than with other processes.
3 More can be done in the way of modifying information on a small-offset machine than on most others.
4 Running costs are lower than with alternative processes other than spirit.
5 The many ways in which masters can be made adds flexibility to the already flexible system.
6 The variety of papers on which the copies can be made is greater than with other duplicating systems.

Naturally with all this flexibility available the machines are more complex and may therefore require more routine servicing. Unlike general purpose offset duplicators they can be operated by virtually anyone, but you will need someone who is able to make minor adjustments from time to time. As with general purpose machines such a person can be trained by the machine manufacturer in a week or two. However, if you do not operate a small-offset print department it is a wise precaution to ensure that servicing of the machines of your choice is prompt and efficient in your area. This requirement will be denied by some manufacturers but none the less it is better to be safe than sorry and the effort is worthwhile, to obtain the many advantages which small-offset has to offer.

What you can do with small-offset

Using the small-offset process you can add, delete or modify the information on the original master, or you can make a secondary master from the first, and this too can be used to make copies on which the information has been modified. There are now even machines on which you can select the area you want to print. Instead of the complete master printing on every copy pressing a button selects the area of the master which will then reproduce automatically. At the time of writing these machines are marketed by only one manufacturer in this country.

It is sometimes said that if you need to carry out all the modifications of which the small-offset process is capable it is the paperwork system which needs overhauling, but it is still cheaper to modify the information on a duplicator than to modify it electronically. The following are the more important ways in which information can be modified:

1 The paper on which some copies are duplicated may be either shorter or narrower than the master, thus deleting information from these areas.

2 The copy paper can be preprinted in any way which fits the system. This includes areas of cross-hatching or reverse (white on black) duplicating. The former deletes information, the latter substitutes the message in reverse type for whatever is underneath. The design has to be carefully worked out, however, to prevent whatever is on the master showing through and confusing the reverse lettering.

3 The copy paper can be punched to indicate a routing or other code.

4 The copy paper can be colour coded or of a different weight or texture also to indicate the use to which certain copies are to be put.

5 The masters can be perforated so that an area is torn off after certain copies have been made and further copies made from the remaining area of the master.

6 A frisket can be attached to a part of the master to mask some of the information on it. This can then be torn off before other copies are made.

7 Two masters can be run together side by side.

8 The inking roller can be modified so that a strip down the master does not print on some copies.

9 A second printing head can be provided on some machines so that two colours can be duplicated at one pass of the paper.

10 A numbering attachment can be added to many small-offset machines to enable forms to be numbered as they are duplicated. Such numbering may be either batch or sequential according to the type of numbering attachment fitted.

11 Masters can be printed with either reproducing or non-reproducing ink either on the same machine or on another machine of a similar type.

12 A rotary collator can be attached to the duplicator so that the copies are automatically collated as they come off the machine.

There are still more ways in which information can be added or deleted or the original otherwise modified but these are more rarely used. Few, if any, installations use all these techniques but the fact that they exist means that the system can be designed to fit the paperwork flow rather than the paperwork flow being adjusted to meet the capabilities of the process. A small-offset duplicator manufacturer will, if the customer insists, adapt the system to an existing flow of paperwork, but it is preferable to redesign the paperwork so as to keep the use of the machine and accessories as simple as possible. The flexibility of the system then enables a paperwork flow to be designed so that it is ideally suited to the task which it is to perform in the overall organization.

Making the master

Reference has already been made to the large number of ways in which a master can be made for an offset systems duplicator, some of these being direct and others indirect. Indirect methods are all based on copying techniques of which the electrostatic one predominates, largely because of its speed and the low cost of short run masters. In fact, several manufacturers base systems on the combination of an electrostatic copier and a systems offset duplicator. Applications vary from customer to customer but the aim of most of these systems is to keep it simple and hence, hopefully, reliable. The basis is invariably that masters are made from originals which are the output of calculating machinery, or have been typed by hand and these are designed so as to produce a document set simply by duplicating on to preprinted paper. The technique described earlier of deleting unwanted information by cross-hatching is also employed. This makes a neat simple system for which the outlay need not be too high, and still leaves the possibility of printing one or more secondary masters either for use on a different machine or the same machine during a different routine. All this can, incidentally, be done on desk-top machines, a convenience when the task is being performed in the same office as the routines which give rise to the need for the copies.

There is, however, also the possibility of making masters directly on the accounting machinery. For this it is necessary to substitute a lithographic ribbon for the normal one and offset masters for the normal stationery set. As with spirit duplication this adaptation should only be made under the supervision of the machine manufacturer. Whether the master is made directly on the machine or indirectly on a copier the concept of duplicating the output of modern electronic accounting machinery, including direct entry computers, has several advantages. Firstly, the checks built into these machines obviate most of the sources of error present when less-sophisticated machinery is used and it is therefore safe to copy a chain of documents from their output. Secondly, the accounting machine itself is relieved of tasks which can be done on the simpler and less costly duplicator. This advantage will become more apparent over the next few years as the full possibilities of these machines are exploited.

CHAPTER 6

Duplicating computer output

Why duplicate computer output? A computer line printer is capable of producing three, four or more copies and you can always run the program again if you require more. Granted that you can run the program as many times as you will but if your sole object in doing so is to produce more copies of the output it would be difficult to find a more expensive way of producing them. After all a computer costs upwards of £30 an hour to run and even one of the small visible-record installations costs up to ten times as much as a duplicator. The alternative is to make do with the number of copies which can be produced on the output printer but this too is no solution to the problem. A computer is designed to produce information in the form of accounting routines, statistics and management reports quickly and accurately. The output printer is merely a peripheral which enables it to communicate this information back to man. Its ability to communicate with several human beings at one time is, as we have seen, both limited and inflexible. Yet it is only when the computer can communicate its findings to all those who need to act on them, quickly and simultaneously, that it becomes the powerful management tool which it is, potentially. The missing link is the duplicator.

Let us assume that the computer's job is finished when it produces the information in human language. Duplication can then take over the job of preparing that information so that it is of maximum use to the maximum number of people. We have already seen that a duplicator is capable of producing a set of documents from one master and it is just as easy to adapt this ability to computer output as to masters produced in any other way. The same basic idea applies whether the computer is being used to produce routine accounting documents or management reports and in either case the computer needs to be programmed with the duplicating ability in mind if the maximum benefit is to be obtained. There is, for example, no benefit in printing out page headings or using expensively printed continuous stationery which has to be changed for every program run if the same result can be obtained by using an overlay on the duplicator or by some other simple device.

However, this principle of programming the computer to take advantage of the ability of the duplicating system can be taken further than merely omitting headings. One application which springs immediately to mind is where the computer is programmed to produce a master for a sales documentation set in the same way that in a smaller unit a similar master may have been produced by hand or on an accounting machine. The various documents which make up the set can then be obtained simply by deleting information from the master during the duplication process. Similar routines can be applied to purchasing, stock and some aspects of production control.

In general terms it is a question of finding out the ability of the duplicating steps in the system and then eliminating these steps from the program. This automatically releases the very much more expensive computing equipment from certain routine steps giving it more time for the complex calculation of which only computers are capable.

There are several duplicating processes which are capable of handling the output from a computer. The one most commonly used today is offset in its systems form but xerographic equipment has recently come on the market with similar capabilities. Diazo duplication has been and still is being used to provide cheap copies of computer output and theoretically spirit duplication could be used, particularly for handling the output from small visible-record machines. Of the four processes, I will deal with systems offset duplication first, both because of its popularity and because it offers such a wide range of alternatives for the duplication of information produced on an output printer.

Small-offset and computer output

The most direct way of dealing with computer output by the small-offset process is to substitute paper plates in the form of continuous stationery for the normal preprinted continuous stationery set in the output printer. To do this the output printer needs to be modified to take a lithographic inking device and if the duplicated copies are to be clear it also needs to be modified so that when the printing heads strike the plate they leave a surface impression. If they leave an indented impression on the paper plate the impression will print hollow due to the offset process not transferring sufficient ink to fill up the depression. It is unfortunately not possible for this to be done on the duplicator as with such over-inking it would be impossible to maintain a correct ink-damping fluid balance. With a correctly set-up output printer this direct method should give sufficiently good quality for internal documentation and for such tasks as preparing routine accounting documents. If you are raising a sales or purchase documentation set from a computer generated master, for instance, you should have no difficulty in producing clearly readable unambiguous documents provided your output printer is correctly adjusted and maintained. This applies equally to a line printer as to the typewriter kind of output provided on many visible-record or direct-entry computors. Further it applies whether the duplication job involves only straightforward repro- duction of the original or obtaining a documentation set from one master using some of the variations already described. Nonetheless you will be working from paper plates and will be limited to com- paratively short runs.

In most documentation systems you will be far from exhausting the life of a paper plate by the time you have taken all the copies you need, but computers these days are capable of producing other kinds of information of which comparatively large numbers of copies are required and/or where the best possible quality copies are required.

Such applications include library lists, voters lists, airline or rail- way tariffs and many similar tasks where the end product is equivalent to a printed list or catalogue. Since it is a valid printing process, offset lithography is ideally suited to this kind of applica- tion. For such applications the quality is much improved if an intermediate process is used to make the plate.

There are a number of such intermediate processes available, two of which are of particular interest because they allow the original to be reduced in size and are capable of keeping up with computer output. One is the photodirect method which, as its name implies, allows a plate to be made from an original without a negative. Consequently, the process is rapid and handling the equipment is well within the capabilities of those who have had no specialist training in photographic work. The other process is electrophotographic and embodies the same advantage of rapid working and comparative ease of operation.

In both cases the equipment used is a specially constructed process camera of the simpler kind originally designed for use with small-offset printing machines. The camera is a fairly expensive piece of apparatus because it must contain a high resolution lens system in order to throw a perfect, or as near as is humanly possible perfect, image on to the plate. Lens systems of this quality cannot at the present time be produced quickly without a fair expenditure of skilled labour, especially when the lens has to cover an area as large as that of a small-offset plate. It is comparatively easy to make a lens which will reflect an image over a small area without distortion, but to produce one where the image is as sharp in the area around the edges of the plate as it is in the centre is the test of a good lens. Apart from this the apparatus, like every process camera, needs to be heavily constructed so as to avoid vibration and to be accurate. It is these characteristics which you are paying for when you buy a good process camera. They are just as necessary when extra equipment is added to the basic camera to enable it to produce plates by one of the rapid methods referred to.

Using either process, masks or overlays can be introduced to add standing information such as company logos, page headings, rules, etc. This not only makes the finished page far easier to read but saves repetitive steps in the computer program, from an economic point of view this latter being the more important. According to the final result needed the artwork for such overlays can be prepared in any of the ways normally used by commercial studios. This gives a very wide latitude and enables a variety of typographical effects to be introduced on the duplicated pages if required. Further, the standing information may be preprinted either in whole or in part, one or more secondary colours being used as desired. In practice this means that, where there is a prestige or advertising element in the documents, it is possible to present them in a form not unlike that obtained by conventional printing methods but still with the advantage of incorporating computer output with a degree of immediacy which would be impossible by other methods.

Before considering other methods of duplicating computer output it is worth noting that more advanced equipment exists for computer typesetting, but this is really the province of the specialist commercial printer.

Xerography and the computer

A comparatively recent development has been the introduction of computer forms printers using the xerographic process. The advantages of these, like other xerographic equipment, are simplicity of operation and virtual silence. However, the particular equipment now on the market reduces computer printout to A4 size or the equivalent in the old sizes and can be used in conjunction with overlays. At the time of writing these overlays have to be positioned by hand but it is only a question of time before a carrier mechanism will be developed. Exactly what form these will take probably even the manufacturer has not finally decided but the mechanics are not too important from the user's point of view. Whatever the arrangement of slides or levers the result will be that by pressing a button or other simple device the operator will be able to select an appropriate overlay.

As I have already pointed out considerable computer capacity can be saved if standing or repetitive information can be transferred from the printout, i.e. from the program to a duplicator. There is, of course, much more to it than this, but if you can transfer an operation from a complex and expensive machine to a less expensive one without loss of time, then a saving is certain to follow. This, of course, applies to all forms of duplication of computer output but it is a fact well worth noting and I make no apology for repeating it.

Xerography, therefore, presents a viable alternative to other duplicating methods. On the machines currently available it is necessary to position the required overlay on the platen, stopping the machine and changing it as necessary, but the carriage of the continuous stationery from the computer output printer is automatic and the machine can be set to produce up to a given number of copies from each original and an automatic collator is available so that these can be stacked in sequence. Speed of operation is in the order of 40 pages a minute, adequate to cope with the majority of computer printers.

As I have already noted computers, which are astonishingly efficient in solving any problem which can be expressed in mathematical terms, are notoriously inefficient when used as duplicators.

On modern high-speed line printers even the third and fourth copy of a multipart set are often difficult to read and multipart sets are by no means cheap. At the time of writing the only user evidence of comparative costs between the use of these sets and xerography comes from the British Aircraft Corporation who have publicly stated that the cost, at a fraction over a penny a page, is very nearly the same. This assumes that all the required copies are produced on the first multipart set, a very big 'if' which favours duplication many times over when a program has to be run a second time merely to produce extra copies. Naturally, this applies equally when some other duplicating method is used.

It is difficult to make a cost comparison between xerography and small-offset. Certainly if you use direct image plates on the output printer and take your copies from these, then small-offset is cheaper than xerography. It is doubtful, however, if the quality will be as good. It may be, with the right operators and well serviced equipment. On the other hand, using an indirect method the quality should be extremely good but the cost will go up. Consequently, it will only compare favourably with xerography if a reasonable number of copies are needed from each page. This number will vary with individual requirements, but if you need more than an average of about 20 copies per page both methods of using small-offset as well as xerography should be carefully costed and the advantages and disadvantages of each weighed before a final decision is taken.

Before leaving the subject of xerography applied to computer output, it is noteworthy that the equipment needed to make masking automatic is in the nature of an extra and that it is likely that, if the manufacturers of xerographic equipment follow their previous policy, it will be usable on the machines which are at present available so there appears to be little danger of having to reorganize computer output printing merely to obtain this advantage.

CHAPTER 7

Uses for microfilm

Microfilm, it has to be admitted, is a rather roundabout way of storing and retrieving information, when compared to hard copy. The usual reason given for filming original documents and then consulting them via a reader is that it saves space. It does, but that is only a part of the story. In fact, microfilm documents take up at most only one-twentieth of the space taken up by the originals.

In some applications this alone is sufficient but with the right indexing microfilm copies can be found more quickly than the originals, especially where they form part of a bulky file. It is quicker to look through a single filing cabinet than through a roomful of filing cabinets.

In other applications the main advantage lies in being able to move a large volume of information around the world quickly and economically. Depending on how much the originals are reduced anything from 60 to 3,000 pages of information can be airmailed from one continent to another for the minimum rate, or at most the next charge up from the minimum.

Another set of applications have security as their primary reason. These include the original application where microfilm copies are deposited in a safe place because the originals are at risk. This, incidentally, is a bonus offered by more active systems. The microfilming of rare documents such as medieval books so that students can study the copies while the originals stay in safe keeping is another use for microfilm where security has high priority.

On the other hand there are a number of applications in business where the original documents pass into the hands of third parties and no copy is made. The microfilming of cheques to prevent fraud is probably the best known application of this kind. It was this application which started modern microfilm as a serious business aid back in 1928. A New York bank, worried by altered cheques, used microfilming as a method of proving the amount for which the cheque was originally drawn. Other applications in the same category include the filming of betting slips, much more difficult to fiddle than a carbon copy, and the filming of library tickets by public and university libraries.

In COM (computer output microfilm) applications the effect of microfilm printout on the data processing system should be the major consideration. Not only is microfilm printout much faster than hard copy printout but it opens up a new set of possibilities for distributing computer processed data at an economical cost. COM has been described as a 'poor man's time-sharing system' and this aspect is discussed later.

What originals can be microfilmed?

The short answer to this question is 'anything which will stand the reduction'. The long answer is much more complex. The standard reduction ratio given in BS4187 Specification for Microfiches is ×18 to ×22 and in BS4210 35 mm Microcopying of Engineering Drawings and Associated Data the maximum recommended reduction is ×30. Even high density fiche on which 200 frames are recorded on a 105 by 148 mm (about 4 by 6 in) sheet of film involves reductions of only about ×40, while ultrafiche such as the PCMI system uses an overall reduction of ×150.

These reduction ratios are considerably less than those involved when a photographer takes a 35 mm picture of a tree, and a 60 ft tree is reduced to 1 inch to nearly fill the frame. The difference is this. When we enlarge the 35 mm slide of the tree on a home screen we enlarge it to something like a twentieth of the size of the original object and we do not expect to be able to identify every leaf. When we enlarge a microfilmed document in a reader we enlarge it to its original size or something near it and unless we can read every character the microfilmed copy will not, in most applications, serve its purpose.

Incidentally there is nothing sacrosanct about being able to read every character. In some applications there may be a good deal of 'small print' which is standard to the type of document microfilmed. If this is lost but the variable information is clearly readable the microcopy may still adequately serve its purpose. As with any other photography what can be microfilmed and retrieved is governed by two basic factors:

1 The quality of the optics in the camera and reader.

2 The grain in the film. This latter imposes an absolute maximum on the amount of reduction an original will stand, depending on the fineness of the grain in the emulsion used.

It would be quite easy here to launch into a technical discussion on resolution but with few exceptions users have to make do with the equipment and materials provided by the manufacturers and work within these limitations. It follows, therefore, that what can be microfilmed successfully depends on :

1 The quality of the equipment used.
2 The quality of the materials used.
3 How carefully the operator uses the facilities at his disposal.

Nevertheless, where the making of the originals is under the control of the organization responsible for having the microfilmed copies made they can economize considerably by making the originals suitable for microfilming.

In practice this means, for example, that originals which may otherwise have to be filmed on to 35 mm can be filmed on to 16 mm or on to one of the high density systems; or, that where it would otherwise be necessary to use a planetary camera the less expensive flow camera technique can be employed.

Users who are new to microfilm often overlook the fact that when something is reduced in height it is reduced proportionally in width, while the space between characters is reduced equally with the reduction in the size of the characters. I know it is obvious when you see it in print but it is easy to overlook when you are designing a form or other original. It follows that a clear sans serif typeface such as Univers will stand more reduction than a fine spiky typeface with exaggerated serifs, such as Old English, to take a rather extreme example.

The other requirement is a good contrast between black lettering on a white background. Ideally the background should be matt as should the ink from which the characters are formed. Unlike the films used in ordinary photography, those used in microfilm are formulated to give maximum contrast between light and dark, with no gradation in between.

This much said it should be remembered that this is a statement of an ideal. The equivalent requirements for engineering drawings are given in BS4210. It is, however, possible to microfilm most originals however unpromising they may at first appear. It does, nevertheless, involve more time, more accuracy and therefore more expensive equipment and in extreme cases using a lower reduction ratio than the one which could have been used if the originals had been designed for microfilming in the first place.

Not only script

When we talk about microfilm we usually think in terms of written or printed records in business or library applications, or of engineering drawings. However, microfilm is a photographic process and anything which can be photographed can be microfilmed. This includes charts, graphs and diagrams of all kinds, including those which are machine generated. Photographic illustrations can be included in microfilmed records, as can photographs taken of three dimensional objects.

This may present some difficulties where it is important to show gradations of tone but even these can usually be overcome. Additionally a colour microfilm has recently been marketed enabling microfilm to be used in applications where colour is a significant part of the record. The only drawback when using colour is that it costs considerably more to duplicate than black and white.

Microfilm is so versatile that it is better to start from the proposition 'What do we need to include in the record?' and then try to find a means of including it and test the cost of doing so, as compared to size-for-size record-keeping of the same information.

The time factor

While microfilm is extremely flexible in what can be recorded, it does involve a time factor longer than size for size copying methods. With exceptions which will be discussed later, conventional microfilm has to be processed and duplicated and this means a delay of anything from two or three hours when the processing is done in-house to normally 24 hours when a bureau is used; this delay being the time between when the information becomes available for filming and the time when it is ready to be examined in a reader.

There are ways of overcoming this delay and undoubtedly 'instant processing' will develop in the future, but at the time of writing it requires special equipment and materials which are only available as dedicated systems. Where 'time is of the essence' one of these systems must be chosen even where another system may otherwise prove more economical.

Updatability

Basically when people talk about an active microfilm system they mean one where the information is intended to be acted upon. Once filmed the information itself is static, it is there in its original sequence for all time. Again there are ways around this. There is one system which enables additional frames to be added to a microfiche at any time after the original record has been recorded. Jacketed film enables a record to be updated by adding frames to the original record contained in the jacket.

What is difficult, however, is to amend a microfilmed record by deleting one or more frames and substituting new ones. By its nature photography is an unamendable process and microfilm offers a permanent record. It is possible to change a frame in a jacket but it is a fiddly business which has to be done by hand. On roll film the only way is to cut out frames from the original film and splice in new ones.

This is more of a handicap in theory than in practice where most applications are concerned but it does mean that when designing a microfilm system it is only practical to film at a stage when all the documents which need to be in sequence are available. For example, where a document set has to be in date order you cannot start filming until the last document in the set is available. You can update but if the documents bearing the earliest date come in last they will be out of sequence. In some applications this is quite acceptable provided it always happens so that you know where to look on the microfilmed record.

Considerations in deciding to microfilm

From this it follows that microfilm is well worth considering if:

1 You need to store essential records in a safe place, such as a bank strongroom where the volume is such that to store all the originals under security conditions would be greater than the cost of microfilming.
2 You need to save valuable office library or drawing office space.

F

3 You need quicker access to information in a bulk file than can be obtained by retrieving the original documents or size-for-size copies.

4 You need to move large volumes of information from one location to another or distribute information to a number of scattered locations at minimal distribution cost.

5 You need to preserve valuable original documents at the same time making the information they contain freely available.

6 You need to publish information of interest to a comparatively small number of readers under conditions such that conventional printing methods would be prohibitively expensive.

CHAPTER 8

Fiche and roll systems

In the present state of development microfilm presupposes some identity of interest between the 'publisher' and user. In this context the word 'publisher' is used in the sense of anyone who makes or orders the microfilm record to be made. Quite obviously the closer the identity of interest the more flexibility you have in choosing a system to meet your requirements. If the information is to be distributed only in-house you have the complete range of formats to choose from. If it is to be distributed only to those with whom you have a long standing contractual relationship you have nearly as much freedom. We are, however, getting to a stage in some areas of both business and higher education where those who are to use your microfilmed information may already have installed equipment for some other purpose. They are then likely to be reluctant to install further equipment simply because your system is incompatible with the one they already have.

For example, British Leyland are already using PCMI to distribute spares listing information to their dealers and distributors. Ford is in the course of installing a similar system. With these two industry leaders using PCMI, accessory manufacturers dealing through the same outlets may find it advantageous to adopt the same system even if their own listings could be accommodated on a few standard 60-frame fiches. (See *Fiche*, page 165.)

This, however, is a purely commercial consideration, where the added cost of using high density ultra-fiche in an application on which a listing may not fill more than half a fiche has to weighed against the cost of providing separate readers for a low-density system.

Fiche versus roll film

Until several years ago roll film was more frequently used than fiche, simply because fiche had not been standardized. Then in 1967 BS4187 was published which followed recommendations made by the International Standards Organisation. It looked as if fiche would settle down to a common format but already technical developments have overtaken this dream of standardization.

According to BS4187 and similar standards published in other countries there were to be two sizes of fiche, one 105 mm by 148 mm on which up to 60 pages could be recorded, and one 75 by 105 mm on which 30 pages of information could be recorded. The single frame size on both is 15 mm by 11·5 mm, the same as that used on 16 mm roll film. A double frame size is also provided for enabling two facing pages to be recorded at one time. There are, of course, a number of further requirements designed to standardize all characteristics of the fiche.

This 'standard' fiche is now in use both in business and academic applications in a number of countries. Unfortunately for standardization some users, particularly in the USA, have found that for some applications it is both more economical and convenient to have more than 60 frames on a fiche. This has led to the introduction of a number of high density systems. By reducing the size of the frame or, put it another way, using a higher reduction ratio a larger number of frames are accommodated on the same size (4 by 6 in or 105 by 148 mm) fiche.

Currently the two reduction ratios used commercially in these high density systems give a maximum of 96 and 200 frames respectively. It is, however, notable that the same size fiche is used and the same basic layout so that to 'retrieve' an image from these high density fiches it is only necessary to change the lens system in the reader. Standardization has come far even if complete standardization has been overtaken by technical progress already.

There are, of course, also the so-called ultrafiche systems of which PCMI and others are available in this country. Since these systems can be considered as a 'micro-microfilm' their use can be thought of as an extension of microfilm usage rather than a direct substitute for comparatively low-density systems. It is for this reason that I discuss them separately.

The advantages of fiche

One of the main advantages of fiche, as compared to roll film, is that the readers are simpler, and therefore cheaper, to manufacture. Further, since there are only 60, 96 or 200 frames on a fiche, compared to 2,400 frames on a standard 100 ft roll of 16 mm film for many applications they allow a much more sensible division of the original information.

In many technical library applications, for instance, the number of frames available on the fiche corresponds much more closely to the number of pages in the original documents. In business applications the smaller number of frames enables a file to be consulted by several users at one time, whereas when roll film is used there is inevitably more chance that under these circumstances there will be more waiting about while the file is being consulted.

This, at least, was the original argument but in practice much depends on the kind of information in the file. What is of greater importance in many business applications is the ability to update information by using the jacketing system mentioned earlier. Further, the immediately updatable system also mentioned previously uses the standard fiche format.

Another advantage which it attributed to fiche is that it is easier to index. The extent to which this is justified again depends to some degree on the application. Each fiche has a 'title space' at the top which is sufficient to hold a title in characters which are large enough to be read by the naked eye and the first frame is usually reserved for indexing.

Mixed fiche

There is one other possibility on fiche, which is to create a fiche on which some frames are of one size and some of another, or where frames are photographed at varying reductions. To be practical such a system must be based on a set of sizes which have a logical progression. This system is being used by the French Centre National de la Recherche Scientifique, Centre de Documentation.

Based on the A paper sizes and using a standard A6 (i.e. 105 by 148 mm or 4 by 6 in) fiche this makes it possible to accommodate all original documents from A0 downwards on the one format. A0 requires one document per fiche, A1 two, A2 four or eight pages of text and so on. Further, when documents are being filmed which will not stand the full reduction you simply use the next size up. Used in conjunction with a reader which is equipped with a lens system capable of suitable enlarging ratios the system provides an extremely flexible potential for microfilming all types of originals within the one basic format.

For example, the same fiche can accommodate an engineering drawing and a complete specification for the equipment it represents. Or, in a library application, illustrations requiring detailed examination can be accommodated at a reduction where none of the detail is lost, while the text is further reduced to a format which is both comfortable to read when retrieved and economic in terms of the number of frames per fiche.

This system is perhaps more advantageous where the originals are not intended for microfilming but it also has considerable potential in any application where the original contains several kinds of information such as text combined with line drawings, maps and photographs. It does, however, need a more sophisticated reader than a constant-size system if it is to be exploited to the full. The user, of course, needs to be able to switch quickly and easily from one enlargement ratio to another, hence a built-in variable lens system is necessary.

As the system stands at present 32, 64 or 112 pages of A4 originals can be accommodated, depending on the amount of reduction they will stand. However, it has the same potential for development to a high density system as other equipment.

Roll film

Roll film, on the other hand, is entirely adequate for large files which are virtually self-indexing. Provided each document is numbered before filming and the reader is equipped with a frame counter no difficulty is experienced in finding individual frames. Roll film is also adequate for all archival records. Since it is not anticipated that these records will have to be consulted, except in an emergency, the likelihood is that if they are needed some kind of printout of the complete record will be required.

This also has a bearing on fiche systems since when making fiche by jacketing roll film it is usual to use a duplicate and the original roll film, treated as archival, is still available for examination as a last resort in cases of doubt. This means in practice that where the information which is normally required is still clearly readable the jacketing system is still acceptable even in situations where other information on the documents can only be read clearly on the original roll film.

Aperture cards

Aperture cards are discussed at greater length in Chapter 15 on Microfilm in the Drawing Office. However, it should be borne in mind that they can if necessary be used for mounting commercial or library documents. Such aperture card systems are advantageous where there is a requirement to re-sort the original documents, even if the size or nature of the originals does not make it necessary to reduce them only to the 35 mm frame size. It is also, of course, possible to obtain aperture cards in which several small formal frames (usually 16 mm) can be mounted.

An aperture card is, of course, an 80-column punched card and 40 colums are still available for punching. Microfilm mounted in this way can, therefore, be mechanically sorted using punch card equipment. Such requirements are most likely to occur in various fields of research.

CHAPTER 9

The microfilm techniques

Microfilm can be thought of as simply a specialist branch of photography in the same way that industrial photography and advertising photography are branches of the same basic technology. Originally both the equipment and materials were shared with other branches of photography but in recent years, like other branches of photography, microfilm has given birth to several techniques which are not used outside this field. This is not because they are unique to microfilm but because the particular features which make them useful in microfilm do not offer sufficient advantages in other photographic fields when compared to available alternatives.

Like every other photographic process, microfilm can be broken down into a number of basic steps:

1 Photographing the originals.
2 Processing the film.
3 Duplicating the processed film.
4 Mounting the duplicated film.
5 Reading (i.e. viewing) the duplicates.
6 Making enlarged hard copies from them (printing).

Photographing the originals
Two kinds of cameras are available for making microfilm records. One is a flow camera which, as its name implies, is used to record large numbers of similar records in a constant stream.

The other is a planetary camera, which records each frame separately. Naturally using a planetary camera results in higher quality images and, other things being equal, it is often possible to obtain a readable record by using a planetary camera; when recording the same document on a flow camera it produces unsatisfactory results.

Flow cameras
To outward appearances a flow camera looks very much like a

rotary office copier. The documents to be microfilmed are presented to a 'copy slot' and as on some copiers automatic feeds for feeding single sheets are obtainable. Alternatively they can be fitted with a continuous stationery feed. On some models it is even possible to change the single sheet feed and substitute a continuous stationery feed in a matter of a few minutes.

Inside, the camera is rather like a movie camera which has been synchronized with a document transport. As the originals pass over the exposure section they pause just long enough to be photographed before being rejected at the rear of the camera. In this way up to 100 ft of documents can be filmed in one minute.

All flow cameras photograph on to roll film, usually 16 mm but sometimes 35 mm stock and in one case 8 mm film is used.

Using an ordinary flow camera it is, of course, possible to record only one side of a document at a time. If you want to record both sides you have to turn the stack of originals over and run it through again. This means that the two sides of any document will be recorded on different parts of the film.

To overcome this a type of flow camera was designed which records both sides of the original side by side. The reverse side of the original is reflected on to the film surface by means of mirrors. Naturally the reduction ratio is greater than when using a conventional flow camera, a point which has to be taken into consideration when designing the originals. This technique is known as 'duplex' recording to distinguish it from 'duo' recording. The latter also uses only half the width of the film for each frame, but the camera records only one side of the original. When the film comes to an end the film transport reverses and the next batch of documents is recorded on the other half of the film width. This latter technique is, therefore, a means of economizing by recording twice the number of originals on the same length of film.

Planetary cameras

These are of two basic types, roll film and fiche, the latter incorporating a step and repeat mechanism. When using a planetary camera each original is mounted individually on the copy board. The camera head, on vertical models which are the majority, is mounted directly above the copyboard and provision is made so that by adjusting the height of the camera from the board any document can be copied at a reduction which will 'fill the frame' if the original has the same format as the film frame.

Size A documents incidentally do have the same format as used on film, regardless of whether 8, 16 or 35 mm stock is used. On most modern planetary cameras exposure setting is automatic and lighting is built in to the camera frame which ensures an even distribution of light over the copyboard. A translucent copyboard illuminated from underneath is provided on some cameras for copying originals with a transparent base, such as engineering drawings on film.

Since the light source is artificial this can be varied in intensity and on some models exposure is fixed (as on a flow camera) and the lighting varied to suit light or dark originals. Many modern planetary cameras, particularly those which are intended for filming business and library documents, are so automatic in operation that the operator only has to watch the focus when the size of the originals is changed. Hence they require only marginally more care than flow cameras on the part of the operator.

Flatbed cameras

In outward appearance these are similar to a flatbed office copier. The documents are placed one at a time, face down, on the copy-bed, which is transparent. The camera itself is situated directly underneath. If you like you can think of it as a planetary camera upside down.

Normally the focus is fixed and the equipment is designed for a 'press button' operation. Such cameras are either special purpose models for filming library tickets, betting slips, etc. or are designed for filming documents which are all the same size or nearly so, for example, A4, quarto or foolscap. Their primary purpose is to film large numbers of originals as simply as possible. They can be considered as an alternative to a flow camera where they are not special purpose models.

Films

The 'standard' film used in all microphotography is a silver halide photographic film formulated to give maximum contrast between light and dark areas (i.e. the 'text' and 'background'). It is very similar to the film used in lithographic platemaking. However, both positive and negative microfilm is available and the resolution varies to some extent depending on the purpose of the film, for high-density fiche will need to have a higher resolution (i.e. finer grain) than one formulated for standard 60-frame fiche or 16 mm roll film. Silver halide photographic films should be used wherever the best quality image is required.

For duplicating diazo films are used. These are films on which the coating contains diazonium salts which, when exposed to heat, are combined with a second material to form an azo dye. Development is virtually instantaneous and duplication on to diazo film is less expensive than when silver halide film is used.

Additionally there is one British fiche system which uses heat development film as the material on to which the originals are photographed in a special step and repeat camera. This camera also incorporates a device known as a 'heat pad' which enables extra frames to be added at any time after a part of a fiche has been used. A complete microfilm system was built round this process with the claim that it would no longer be necessary to keep any original documents. Although this has many attractions it would mean equipping every work station with a microfiche reader and providing reader/printers for those documents handled by anyone who did not have reading facilities. The equipment which was designed in Britain is marketed by Birch-Caps Equipment Ltd. There are also several proprietary vesicular (instantaneous developing) films now reaching the market suitable for both photographing originals and duplicating.

An alternative system which also allows microfilm to be developed in under a minute is the dry silver process marketed by the 3M company. Up to the time of writing this has been applied to equipment designed primarily for engineering drawings as it is more suitable for 35 mm than for the smaller formats.

Monobath films are a modification of the normal photographic formula which enables the film to be developed and fixed in a single bath. Equipment designed for this type of film usually makes provision for guillotining one or more frames from the roll and developing these separately before the rest of the roll has been exposed. This is probably the main advantage of monobath films which, if not quite up to the resolution standards of the best standard films, are entirely adequate for most purposes.

Film processing

Film processing can be undertaken in-house either in a normal photographic darkroom (this is usual only where the company already has a fully equipped photographic unit on the premises) or automatic processors are available which are capable of handling all grades of standard microfilm stock. Most of these automatic processors are 'daylight loading' (i.e. no darkroom is needed). They do, however, require an operator who is sufficiently knowledgeable to keep the tanks filled according to the instructions and to clean them out regularly, which involves simple disassembly and re-assembly of some of the parts.

Monobath films are also normally developed in an automatic processor but as this requires only one bath the process is considerably simplified and speeded up.

Materials such as dry silver, diazo and other vesicular films are normally developed in a camera/processor or duplicator/processor. Most such equipment is suitable only for use with one type of material. You can sometimes 'get away' with using equipment designed for one exclusive trade process when using materials from another but if you do this you do so on your own responsibility.

Duplicating

As previously mentioned microfilm can be duplicated either on to silver halide or vesicular film. Automatic equipment is available for both processes. Vesicular duplication does not usually give quite the same high quality image as silver halide and the latter should be used where the film is to be processed further (see 'jacketing' below).

Mounting

Depending on the microform in which the film is to be used there are a number of ways in which the 'active' duplicates can be mounted:

1 ROLL FILM   This can either be spooled or, more frequently these days, it is mounted in a cassette. However, where this is done a reader or reader/printer designed to handle cassetted film must be used.

2 JACKETING   Roll film can be cut and mounted in trans-

parent holders known as jackets. The actual jacketing can be done automatically in a mounter designed for the purpose. Usually the jackets have the same format as a fiche, i.e. several rows of pockets running across the longer side of the jacket so that when they are filled the result is similar to a fiche.

The advantage of the system is that if all the pockets are not filled frames can be added at a later date to update the record. It is usual, when using jacketing systems, to duplicate the jacketed record on to fiches and use these as the 'active' record. This results in some deterioration of quality compared to the original microfilm but if the originals are clear and the 'text' bold no difficulties should be experienced.

Although it is usual for jackets to accommodate single frames all of the same size, it is possible to mix frame sizes on a jacketing system; for example, two rows of 16 mm frames and one of 35 mm frames. This should only be done if strictly necessary, e.g. to accommodate engineering drawing with related specifications, as it will entail some hand filling of the jackets which is a time-consuming operation. It is then usual to refilm the jackets and distribute fiche copies of the jacketed film originals.

3 APERTURE CARDS   These are usually used to mount single 35 mm frames for drawing office applications. Equally they can be used in this way for oversize documents of other kinds. As previously mentioned an aperture card is a standard 80-column punched card and 40 columns are still available for sorting automatically if needed. Usually, however, blank cards are used for sorting and the aperture cards are then 'pulled' from the file.

There are also a number of individual aperture cards available to accommodate from one to sixteen frames, either 16 mm format or mixed 16 mm and 35 mm. Except for some specialist applications these would appear to have few advantages compared to either single frame 35 mm aperture cards or fiche, depending on the application.

4 LARGE FORMAT FILM   Large format film, either 70 mm or 105 mm, is usually filed unmounted, each frame being protected by a cellophane envelope.

RETRIEVING THE RECORD

Readers

Microfilm readers vary considerably in price, type and in their ability to accommodate one or more microforms. The least expensive readers are those made to read standard 30 image or 60 image fiche. Most are of the back-projection type, the image being thrown up on a translucent screen which typically measures about 8½ by 11 in. This gives approximately size for size reproduction of A4 originals.

The essential elements in any reader are the lens system, the lamp, the reading screen and the film transport. It is because in a fiche reader the film transport is a simple mechanical device which enables the fiche to be moved in the x and y axis that they are less expensive than roll film readers. Provision should also be made to rotate the image through 90 degrees, as in standard filming practice documents which have the text centred on the long side are not turned. If the reader is to be used constantly without eyestrain the lens system must be arranged so that it throws a sharp image on the screen and the screen should be coated in such a way that there is no glare when it is illuminated. If you want to see if the lens system is good examine the outer edges of the screen for signs of 'fuzziness'. It is easy enough to make a lens which gives a sharp focus over a small area but not so easy to make one which throws a sharp image over the entire screen.

Apart from the least expensive models many readers are designed to accommodate several film transports and have interchangeable lenses. Some are designed to accommodate both fiches and aperture cards but it is usually necessary to change the film holder by hand. Others include a roll film transport but on models primarily designed for reading fiche this is usually of the type which has to be cranked backwards and forwards by hand.

Certainly these refinements tend to add to the price of readers. Whether you need them or not depends on your application. If you receive microfilm from outside your own organization the more flexible the reader in respect of the number of microforms it will accommodate the better, but if you operate an 'in-house' system you will naturally standardize on one microform.

Another choice, so far as fiche readers is concerned, is between desk-top and portable models. The latter are obviously necessary if you want to read microfilm while you are travelling about. On the other hand desk-top readers are more suitable for continual reading as opposed to referring to the microform records. Even so, much depends on the quality of the individual reader.

There are a few readers available which are of the forward projection type, although one portable model can be used either way depending on how you set it up. With forward projection models you simply use your blotting pad or any other suitable surface as the screen. The size of the 'picture' then depends on the distance of the surface from the lens.

Roll film readers, on the other hand, are invariably motorized if intended for continuous use. The same remarks apply to the general construction as applied to fiche readers. However, since there are up to 2,400 frames on a standard 100 ft 16 mm roll—the use of 35 mm film in roll form is rare except in newspaper publishing— they are equipped with facilities for motoring the film forwards and backwards usually at two speeds. They should also be equipped with a frame counter as at the higher speed it is impossible to see anything on the film.

The usual way of using a roll film reader is to motor it at the higher speed to somewhere near the required frame watching the counter, and then edge it to the required frame at the slower speed. An added refinement is a self-centring device which automatically centres the selected frame. In this way any desired frame can be located in seconds provided the record has been adequately indexed. Simpler roll film readers on which the film is wound by hand are adequate for occasional use and fiche readers are frequently offered with adaptors of this type.

Aperture cards

The simplest aperture card readers are designed to mount 35 mm aperture card mounted film, and nothing else. These are adequate for use in a drawing office where the complete microfilm record is stored in this way. Most have a lens system capable of giving two or three magnification ratios so that parts of a drawing can be 'blown up' to fill the entire screen when required.

This is a particularly useful feature on readers used in a drawing office as it frequently obviates the necessity for a draughtsman to ask for a hard copy. For instance, a draughtsman making a modification can comfortably work from a reader displaying the part of the drawing he is working on at about full size, whereas he may find the complete drawing displayed on an average sized screen a little small.

In practice many readers are available with both fiche and aperture card mounting attachments. You do not necessarily have to buy both as one or the other is usually an 'optional extra'. It is, however, necessary to ensure that the enlargement ratios available using the lens system offered are suitable for the kind of originals you want to read.

Reader/printers

The reading part of a reader/printer is exactly the same as a reader pure and simple, and with few exceptions reader/printers can be used as readers when no printout is required. The printout is simply a hard copy of the frame which is presented on the screen. Any of the office copying processes for which projection speed materials are available can be used. Today, because of its convenience, the electrostatic method is most frequently used.

One refinement found on, at the time of writing, two electrostatic reader/printers, is the ability to produce a positive or negative hard copy from either positive or negative film. Other reader/printers use the diffusion transfer process and even in one case diazo.

For business documentation and library work a reader with a capability of printing out copies of about A4 size is adequate. Such copies should be confined to those needed by people who do not have reading facilities if costs are to be kept down. There are exceptions but in most applications the production of unnecessary hard copies defeats the object of microfilming in the first place.

If you need to study a document in detail then it may be necessary to make a hard copy. Nevertheless, the smaller the number of hard copies required the more economically you can operate the system.

In addition to reader/printers of the kind described a number of special purpose equipments are available. One such equipment, made by ITEK, prints out on to small-offset paper plates and is intended for use in micro-publishing, enabling hard copies to be offered as an alternative to microfilm.

Alternatively there is a xerographic equipment available which

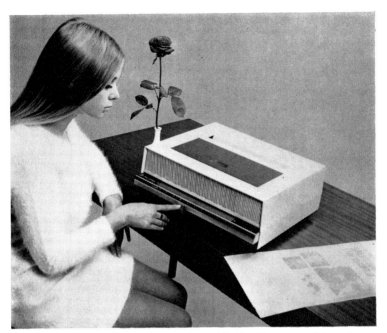

Plate 1. Heat transfer copiers such as this Fordifax are quick and easy to use, although they have limitations for straightforward copying. In addition they can be used to make inexpensive masters for spirit and stencil duplication as well as overhead projection transparencies.

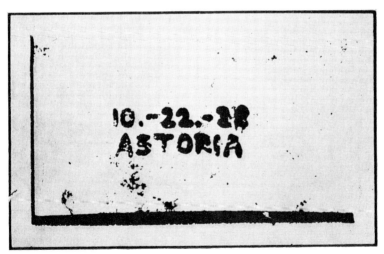

Plate 2. A historical curiosity, the first xerographic print made by Chester F. Carlson and Otto Kornei on October 22, 1938 in Astoria, New York.

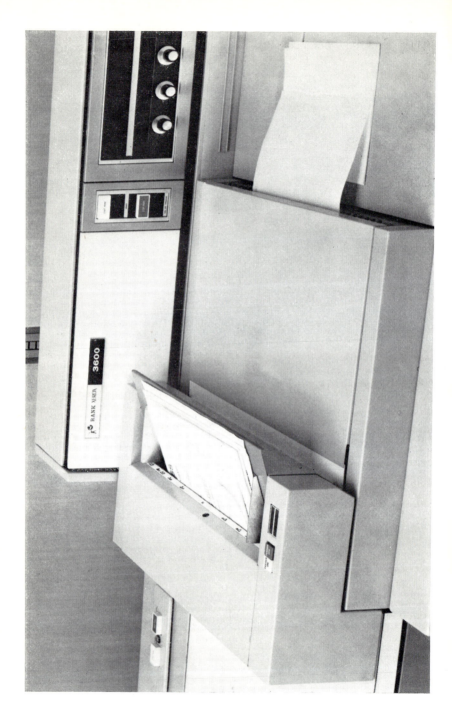

Plate 4. In contrast to a fiche camera a flow camera films onto roll film at high speeds, virtually as fast as the documents can be fed into the camera, being therefore ideal for high volume work. Many such cameras are available with automatic feed devices or with special feeds for continuous stationery.

Opposite Plate 3. Automatic feeds are particularly useful for systems work.

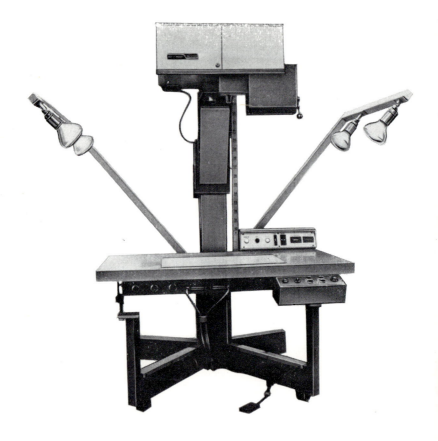

Plate 5. A typical microfiche camera. The camera head is designed to 'step across' to the next position on the fiche as each new original is photographed.

prints out from microfilm on to plain paper and has a similar function. However, we are now getting into the realm of micro-publishing, which is discussed separately. Printout of engineering drawings is also discussed separately in the chapter devoted to engineering microfilm (page 117).

CHAPTER 10

Bureau services

The majority of microfilm users do not do their processing and duplicating in-house, and many do not possess a microfilm camera. All this can be done by one of the bureau services set up for this purpose. In fact, unless you have a large requirement it is uneconomic to do your own microfilming in-house.

Exactly where to draw the line depends on whether running costs are of primary importance or whether time is particularly important. The turnaround time for standard fiche or roll film is in the order of 24 hours. You may, however, be able to make individual arrangements to shorten this period if you have a regular job which a bureau will do on a priority basis. Even in discussing microfilm, time is relative, depending on the tempo of your business. Whereas a minimum of 24 hours lapse between the time the information becomes available and the time when the microform is ready for reading may be intolerably long in a few applications, it is a very considerable improvement over any reasonable alternative in others.

Starting with a bureau

Whatever your initial application it is advisable to start by using a bureau service. In this way, even if your system is large enough to justify installing your own equipment, you can test whether your application will live up to its initial promise.

Ideally the documents should be designed with microfilming in mind but even where they originate in-house it may be necessary to use existing forms during the changeover period. In such cases it is possible that while the initial documents will need a planetary camera to record the information clearly, redesigned documents may make it possible to use the faster flow camera.

Some bureau services are operated by equipment manufacturers. These usually produce high-quality work but it should be borne in mind that their eventual aim is to sell equipment and materials of that particular make. Usually, but not always, they are based on the microforms which make up the equipment range sold by the same company. Some, however, offer services for which they do not sell equipment.

The point here is that you have to shop around to find a service which suits your requirements and, if you are thinking of installing your own equipment at a later date, preferably one operated either by an independent bureau or one which sells the kind of equipment you are thinking of buying.

In fact, it is never necessary to buy any equipment, not even a reader, as microfilm, like other equipment, can be hired either directly from a manufacturer or through one of the finance houses which specializes in this type of transaction. In some ways this latter course is preferable as microfilm equipment is developing rapidly and what is the best buy this year for your application may easily be superseded next year or the year after by equipment from a different manufacturer.

The different services offered

Most microfilm service houses offer to film original documents on to roll film and process it. From here onwards the variety of services offered diverges considerably. Virtually all the microforms discussed in Chapter 9 are obtainable from a bureau and additionally some bureaux offer to microfilm on to high density fiches, i.e. 200 frames to a 4 by 6 inch fiche. Additionally there are bureaux which will film the originals on your own premises if you do not want the documents to leave them.

Security

There is certainly no more risk of information getting into wrong hands when documents are sent to a microfilm bureau than when they are sent to a computer bureau. Once the documents have been filmed they cannot be read by the unaided eye and employees have in any case no time to examine customers' documents even if they had the inclination. If you have any doubts as to the security of information you should discuss this frankly with any bureau you may contemplate using before sending any work to them.

Security arrangements vary from one bureau to another and in any case it is a wise precaution to ensure that any documents entrusted to a bureau will be adequately insured against fire, natural hazard, accidental damage and theft while they are in the care of a third party.

CHAPTER 11

Microfilm publishing

In microfilm the term publishing covers a wide range of activities. Some information published on microfilm is available to anyone who cares to pay for it, i.e. is published in the usual commercial sense, while other information is available only to a restricted group of subscribers with a common interest. In one sense all microfilm publishing is restricted since the average person does not have the facilities for reading microforms.

This itself has considerably restricted the kind of information which it is worth a publisher making available on microfilm. Certain types of work which are suitable for microform publication remain unpublished simply because insufficient people who would be interested in them have access to readers while, at the same time, there is a tendency among some organizations to hold back from installing reading facilities until the publications become available.

This situation is partly due to lack of standardization, despite the efforts which have been made in this direction over the past decade. At least we have reached a stage where most fiche is 105 by 148 mm (A6, 4 by 6 in) with 75 by 125 mm (about 3 by 5 in) as a less-used alternative. Roll film used in micropublishing is either 35 mm or 16 mm. In addition there is a considerable volume of information on PCMI which I will discuss later in the chapter.

It is possible that unless there is a considerable breakthrough in technology the present set-up will be with us for some time, although the development of 'superfiche' with about 200 images on a 105 by 148 mm fiche further complicates the issue. After all a considerable amount of the information on microfilm comes from American publishers, and although no American publishers are using 'superfiche' exclusively, and as yet only a few are offering it as alternative to 'standard' fiche, in this context its development needs to be watched.

Newspapers on microfilm

Back numbers of newspapers are a natural subject for microfilm publication. Because of the large page size and small print it is impractical to reduce them below the 35 mm and by far the greatest majority are on 35 mm roll film. The original newspapers being frail and difficult to handle, as well as bulky to store, microfilm offers an ideal solution to the distribution of the vast store of historical material which is available in this form. Further, since a newspaper publisher's main interest is in the present there is seldom any objection to publication on microfilm, even when the material is of comparatively recent date. On the contrary, most publishers are pleased to be relieved of the responsibility for having to keep back numbers. This condition applies also in the related periodical field.

Consequently there are a wealth of newspapers and periodicals from all over the world available as microfilm copies. Many of these date back to the time when the newspaper was first published and many newspapers and periodicals which have gone out of circulation are included in various publishers' lists.

As far as current publications are concerned, these are usually only available to those who subscribe to the normal paper edition. The assumption is that the normal paper edition will be used until it is worn out after which it will be replaced by the microfilm edition.

However, there are a number of periodicals which are either not copyright or on which the copyright has lapsed, or which have been exempted by arrangement with the normal publisher and the organization making the microfilm edition. Such journals may be obtained on microfilm regardless of whether the purchaser is a subscriber to the paper edition or not.

Books on microfilm

Out-of-print books are another major field of microfilm publishing. These are available either on roll film or fiche. As with periodical publishing the largest source of out-of-print book publishing is the USA where some twenty-odd publishers issue books on microform on a purely commercial basis, not counting a number of small publishers of purely local interest. Some of these books are also available in hard copy form, the master microfilm having been 'reprinted' by using the xerographic process or some similar copier/duplicator process.

Although many microform publications are of purely American interest the leading publishers include among their titles a fair proportion of works by British authors. In addition a number of European works are included either in translation or in the original language.

The advantage of microform books, whether actually in microform or in hard copy printout, for out-of-print works, is that the price is related to the cost of production rather than to any rarity value the original edition may have on the secondhand book market. Consequently most works are considerably less expensive on microfilm and where listed can be obtained much more quickly than if a search has to be made through the book trade. Generally speaking, works in microform are approximately one-third of the price of the same works in hard copy form where the hard copy has been made from a microfilm master.

A good basic guide to all that is available from American publishers, both commercial and constitutional, is *Guide to Microforms in Print*, published by Microcard Editions, 901, 26th Street, NW Washington DC 20037 (Price $6). Microcard Editions is a part of The National Cash Register organization. I understand, however, that copies of the guide are not normally sold by the British affiliate company.

Another guide to microforms which is well worth having is that published by University Microfilms. This is in two parts; one entitled *Serials on Microfilm* and the other *Books on Demand*. In the former some 6,000 periodicals are listed, while the latter lists over 35,000 titles. All these publications, which include many British titles, some dating back to the seventeenth century and beyond, are available from University Microfilms Ltd, in London.

One of the chief sources of information on microform publication in Europe is the Microfiche Foundation, Delft, Holland, which publishes a general guide to literature in microfiche form which is commercially available. Unfortunately it does not also cover information on roll film. Another particularly useful list published by the Foundation is that giving the names and addresses of libraries which will make information available on microfiche as requested. Other information given includes the general fields in which information is available, the size of fiche used and the price per fiche. Unfortunately the size of fiche in use at various centres varies considerably. Nonetheless a very useful amount of information is available from the sources listed to any library which has a reasonably versatile reader. Collectively the sources listed cover all the humanities and most sciences and include among them the collections of several major European libraries.

Books in English

Books in English is one of the most interesting developments in the microfilm field to date. At the time of writing it is still in the field trial stage. It is a bi-monthly bibliography compiled from information held in date form by both the British National Bibliography and the American Library of Congress. As such it is said to be the most comprehensive bibliography of English language books available.

For over two years now all the books bought or received by the Library of Congress and by the British National Bibliography have been catalogued on a computer. Magnetic tapes of these machine held catalogues have been made available to those libraries large enough to have access to a computer capable of handling these tapes and running computer programs which enable them to draw off those catalogue details which they require.

The BNB itself uses these tapes to produce the weekly lists which it has published for the past 22 years, printing out on to a computer controlled phototypesetter. With titles being added at a rate of some 700 a week to the BNB list and 1,800 a week to the combined Library of Congress–BNB listings, the information available is rapidly reaching nightmarish proportions.

Quite obviously some means of accessing this information is needed which does not involve libraries either in hiring expensive computer time or accumulating bulky masses of hard copy. With this bulk of information even 'standard' fiche would involve something like 120 fiches to accommodate the listing of six months' accumulation of titles and in an 'add on' index several fiches would be added each month.

Consequently the PCMI system was chosen. For the initial issue the period January to June 1970 was selected. The bibliography for this period contains about 150,000 entries relating to 40,000 English language works and is contained on three PCMI fiches, each 4 by 6 in (i.e. A6). The bibliography is being updated every six weeks on an accumulative basis so that the last issue of each year will be a definitive set of all books published during the year.

The complete production cycle from the time that the Library of Congress and BNB information is merged to the posting of the duplicate fiches takes fifteen days. This involves first converting the Library of Congress records to full compatibility with BNB machine readable catalogue (MARC) records, selecting, sequencing and formatting the magnetic tapes, another computer operation, and finally merging the two re-formatted tapes.

Even then the tapes have to be converted from 9 track to 7 track mode before they can be read by the Stromberg Datagraphix 4060 computer output microfilm unit. This converts the digital information into alphanumeric form and displays it on the face of a cathode ray tube. Each image is then filmed on to 35 mm microfilm at a transfer rate of one frame a second. This 35 mm microfilm is then re-filmed at a 10 times linear reduction on to a photochromic image (PCMI) plate. From this contact copies are made and sealed by a lamination process. These copies are distributed to those libraries taking part in the initial field trials.

Final costs of this service are not available at the time of writing but it seems fairly certain that each library subscribing to the service will pay between £50 and £100 a year. In addition they can hire a reader which can be rented for about £70 a year depending on the maintenance contract. The bibliography is in two main sections:

1 A classified section arranged by the Dewey Decimal Classification.
2 An alphabetical section with entries under author, title and added entries.

A location index is provided to broad areas in both sections, the location being expressed as a three-part reference representing fiche, row and column number. Despite the high density of the fiche used it should be possible to locate any required entry in less than a minute, provided the user is familiar with the indexing system.

Other PCMI publishing projects

Apart from *Books in English* it is probable that libraries which take advantage of this service will be able to use their readers to access reference information, either by the time the *Books in English* service becomes generally available, or within a reasonable time. Such a service is at present being planned by the NCR organization in this country but plans are not yet sufficiently advanced for any details to be available.

Much will depend on copyright and on the attitude of British publishers, but I understand that the proposed service will basically follow the one already available in the USA. British libraries can, of course, subscribe to the American service and a few already do.

Known as the PCMI Information System initial collections, each consisting of 100 fiches, representing about 700 books, are available in each of five subject areas:

American civilisation
Literature—humanities
Science and technology
Social science
Government publications

While the first and last of these collections are of interest primarily to scholars who have made North America their subject the other three collections are far more international in interest and include a fair proportion of works by British and other European authors. Subscription to these American collections costs the equivalent of under 50p per volume. Whether it will be possible to keep the cost of an equivalent British information service down to this level remains to be seen. All that can be said with certainty at the present time is that once the necessary negotiations can be completed it should be possible to make a comparable service available to British scholars at a very reasonable cost.

The twain shall meet

One of the difficulties which is at present being experienced in microfilm publishing is that there is insufficient exchange of information between those who have the financial and technical resources to produce the publications and those who will use them.

As with any other new medium, publishers of traditional hard copy books tend to see microfilm as a threat to their business rather than as another means of publishing in which they can participate profitably. Rather than offer a reference work either in microfilm or in hard copy at a price differential which reflects the difference in production costs they prefer to ignore microfilm even where they know they are losing sales because of the high cost of their product. In the meantime publishers on microfilm are forced back to the publication of works which are out of copyright. This means that the user sees microfilm as a way of obtaining historical information, or in the scientific field, abstracts.

Quite obviously publishers of serious reference works cannot be expected to consent to microfilm publication of their works unless they are convinced it will be to their advantage. On the other hand it would be prohibitively expensive for publishers on microfilm to assemble the expertise necessary to produce independent authoritative works. So, for the moment, we are at an impasse.

This can be broken in one of two ways. Either the publisher of an authoritative work will work out a packaged deal whereby his readers are offered the hire of microfilm readers as a part of the subscription to a microfilm edition of his work. If the terms are right and subscribers benefit, for example, by a more up-to-date service and/or multiple subscriptions at a considerably reduced price after the first, the profitability of this kind of publishing will have been proved. Otherwise a publisher of an authoritative reference work may price himself out of the market and the pieces will be picked up by a microfilm publisher with the same end result.

Servicing manuals on microfilm

A number of applications have been found for publishing in the restricted sense of distributing spares and servicing information to agents and others with whom the 'publisher' has a contractual arrangement. Some of the earlier applications were in the aircraft industry and it is hardly surprising that these are on roll film, as they date back to the period before fiche became at least partly rationalized. The advantages of microforms, roll film or fiche for this kind of application lie in the ease with which the microform can be transported, compared to hard copy, plus the fact that it is sufficiently cheap to make the reissue of a complete list or manual feasible when amendments are needed.

Everyone who has been involved with engineering servicing knows the problems of amendments. However careful the publisher may be in issuing these together with explicit instructions on where they should be inserted inevitably some users will file them incorrectly or even put them on one side. The consequences of wrong spares being ordered and sub-assemblies being incorrectly re-assembled can often be traced to this alone. This is particularly true of the motor industry and of other durable consumable industries. The manufacturer is blamed by the customer for poor service but it can equally well be the fault of the agent.

It was to overcome the prohibitively high cost of distributing spare parts lists running into several thousand pages on a world-wide basis, plus the other disadvantages mentioned, that BLMC Service Department has turned to PCMI fiche. In hard copy form their spare parts lists comprised 72 volumes, something in the order of 18,000 pages, which were being constantly updated. These were distributed to some 2,000 dealers and distributors in over 100 countries. The changeover to microform is now completed and the entire catalogue is now recorded on six fiches. In the meantime each of their 2,000 odd dealers and distributors have been provided with readers.

Using the standard pricing structure which NCR will quote to any prospective customer it is possible to obtain a near approximation of the cost of this operation as follows:

A master PCMI fiche costs from £320 to £350; duplicates cost 50p each. To distribute by airmail it is safe to assume that six fiches weigh under 4 oz including packing; they can, therefore, be sent to virtually any destination in the world at the minimum air parcel rate, or by letter post for very little more.

Six master fiche @ £350	£2,100
6 × 2,000 copy fiche @ 50p	£6,000
	£8,100
Postage 2,000 × 30p	600
	£8,700

Cost per page approximately 0·05p including distribution. This compares with the cost of standard fiche where the originals are made on a step and repeat camera as follows:

Cost of each master fiche £1·50
Cost of each copy fiche 5p

The cost of both master fiche varies between £1·25 to £1·50 and that of copy fiche from 5p to 7½p approximately

$$\frac{18,000}{60} \text{ pages} = 300 \text{ fiche @ £1·50} \qquad \text{£450}$$

$$300 \times 2,000 \text{ copy fiche @ 5p each} \qquad \underline{\text{£18,000}}$$

$$\underline{\text{£18,450}}$$

So many variations are possible that it is difficult to compare this to hard copy publishing. The following is based on small-offset print bought from a commercial printer using a combined size-for-size electrostatic platemaker and single page per sheet small-offset printing:

Cost of printing 1,000 copies 16 pages double sided   £84

$$\text{Cost of printing 18,000 pages} = \frac{18,000}{32} \times 84 = \text{£37,750}$$

It can be argued that this cost could be reduced considerably if the printing is done in-house, but it would be a very efficient internal print department which can halve print costs when over-heads are properly accounted for.

Another interesting comparison is to assume that 100 copies only are required of 3,000 pages of original material. The comparison then comes out something like the following:

1 PCMI master fiche	£350
100 copy fiche @ 50p	£50
	£400
$\frac{3,000}{60} = 50$ master fiche @ £1·50	£75
$50 \times 100 = 5,000$ copy fiche @ 5p	£250
	£325
50 master fiche @ £1·50	£75
5,000 copy fiche @ 7½p	£375
	£450

This is the approximate break-even point depending on the price you have to pay per copy for 'standard' fiche. On this basis it is a reasonable claim that PCMI is more economical than 60-frame fiche where more than 100 copies are required from somewhere between 2,500 and 3,000 pages of original information. If most of these copies have to be distributed by air mail this could just tip the balance.

There may be overriding policy reasons why PCMI should be

considered as an alternative to either standard 60-frame fiche or 200-frame fiche even where it is not justified on strictly economic grounds. For example, now that PCMI has been adopted by BLMC Service Department, and more recently by Ford of Europe, it might be justified from a marketing point of view by other companies distributing through the same outlets even where the volume of information they wish to distribute amounts only to a few hundred pages. I am merely citing this as a example of where marketing considerations might override the economics of production. It was, incidentally, found that a garage storesman could locate a part in 20 seconds when it was in frequent demand and in 40 seconds when it was seldom required.

In this case the reader would be mounted on a desk or work bench, but for reading standard density fiche, or aperture cards, it is possible to obtain portable readers. At least one company is equiping all its service engineers with portable readers as a part of their servicing kit. This would appear to be a particularly advantageous application where service is normally undertaken on customers' premises. Such readers are not expensive and can be built into a case to avoid accidental damage. Except where servicing the simplest machinery, no service engineer can be expected to carry every modification in his head and the normal diazo copies of engineering drawings and duplicated manuals do not last long under these conditions, besides being expensive to produce. Double page readers are also extremely useful in this type of application since they enable two adjacent frames to be seen simultaneously; thus a diagram and the text relating to it can be seen together.

Up to now I have compared PCMI to fiche made on a step and repeat camera, but this is one application where advantage can be taken of the lower cost of preparing the masters on a flow camera and then jacketing. The changeover period from a hard copy system apart, all information can be prepared to the required microfilm standards and consequently will be clearly readable.

General criteria for publishing applications

As previously mentioned there has to be some special relationship between the 'publisher' and user for the distribution of information on microform to be a practical proposition. Either the publisher must himself be willing to provide a suitable reader, or the user must have a sufficient incentive to acquire his own. In business this usually implies a contractual relationship, while in the field of research and higher education the incentive so far has been that the information cannot be easily obtained otherwise or is very much more expensive.

Views as to whether microforms are suitable for 'solid' reading as opposed to reference vary. Much depends on circumstances. Most people these days watch television for long periods without discomfort, but in this they are only watching a general impression. It is one thing to watch a picture and another to apply the degree of concentration necessary to read. Much must depend on the age of the reader. A student, anxious to complete a thesis will be more inclined to put up with any inconvenience he may find in reading from microfilm than a mature adult who is pursuing a course of instruction merely out of interest. Likewise in business an executive will need to be convinced that he will gain from reading the microfilm. He will not, for example, be much inclined to take the trouble to mount sales literature in a reader if he can obtain the same information in the more familiar hard copy form.

Typography and lack of colour undoubtedly play a part in the association of microfilm with useful rather than pleasurable applications. It has to be admitted that where material is prepared for microfilm the ideal typography from a technical viewpoint is utilitarian rather than elegant, while the high cost of duplicating colour precludes the use of colour illustrations in the majority of applications.

This much said, microfilm has considerable advantages over hard copy information distribution where:

1 A considerable volume of information needs to be distributed at one time.
2 The number of copies required is comparatively low, i.e. within the range of 'short run' duplication.
3 The copies need to be distributed over a wide geographical area.

In these circumstances the cost of microform distribution will be considerably lower than when any other means are used. All three criteria are not necessary to every successful application and it is worthwhile investigating any situation where one or more apply.

The PASCAL Project

The PASCAL project operated by the French National Centre of Scientific Research analyses some 9,000 journals, conference proceedings, etc. to produce about half a million abstracts a year. The abstracts are divided into 30 monthly sections covering different disciplines in exact science, technology and biology. Subscribers can subscribe to one or more sections and the information is distributed either in microfiche or hard copy form. Various other services are available including a rapid dissemination service, retrospective searching and personal profile services agreed with users.

Insofar as the information is computer stored and sorted this service bears a resemblance to *Books in English* although it is designed for a different readership.

CHAPTER 12

Computer output microfilm

Computer output microfilm or (COM), was originally applied to techniques which translated information held in digital form into alphanumeric form and transferred them to microfilm without using a hard copy printout as an intermediate stage. Recently, particularly in America, some people have tended to refer to any equipment suitable for microfilming information printed on a computer line printer as COM. This is an obvious attempt to cash in on the more sophisticated process and can be confusing.

While microfilming from computer output print, which to all intents and purposes is the same as microfilming any other continuous stationery, has a number of useful applications, it does nothing to cut the cost of computer printout which is one of the objectives of COM. A COM equipment, in the original sense of the word, is made up of two pieces of equipment:

1 An electronic device which converts digital characters (usually held on magnetic tape) into alphanumeric and displays them temporarily on a cathoderay screen.
2 A camera which photographs the face of the cathode ray screen and is synchronized with the display so that every time the display changes the camera moves on one frame and photographs the next display.

There are variations on this basic technique which I will discuss later.

The technique has the following advantages:

1 The filming takes place at least five times faster than the fastest line printer. Hence, if the filming is done on-line, i.e. with the output connected directly to the central processor, the speed of outputting the information more nearly corresponds to the electronic speeds of working in the computer. Thus expensive computer time is saved.
2 The microfilm can be treated in any way which will ensure

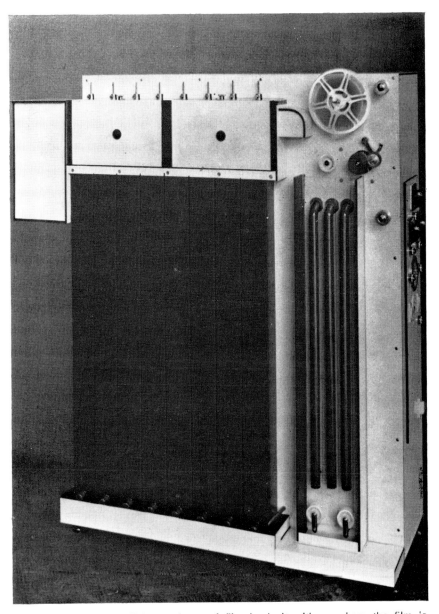

Plate 6. Where a sufficient volume of film is dealt with or where the film is required within an hour or so of being recorded an automatic processor, as here, enables a user to process his own films with a minimum of darkroom facilities and without having to employ a skilled operator.

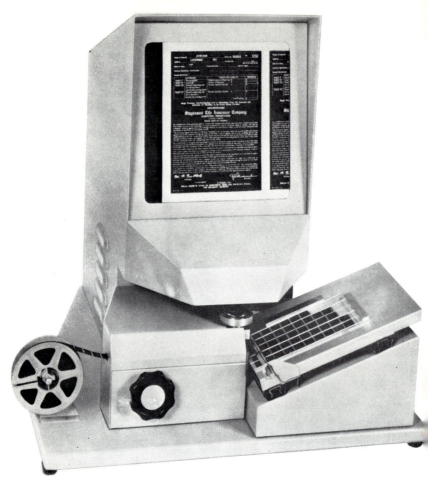

Plate 7. A reader/filler for microfilm jackets. Jacketed microfilm is simply roll film cut into strips and inserted into a plastic holder in a format similar to a fiche.

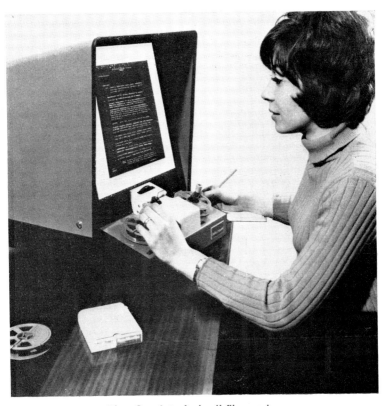

Plate 8.   A typical roll film reader.

Plate 9.   A typical microfiche reader.

an optimum distribution of information, i.e. it can be duplicated and distributed in whole or in part to as many people as
require the information. Further it can be retrieved either in
whole or in part in hard copy form for those who do not have
reading facilities.

It is for this reason that COM has been referred to as a 'poor man's
time-sharing system'. As with everything else speed costs money
and in designing COM based information systems due regard must
be taken of the time taken to process and duplicate the film.

The Cost of COM

A COM equipment is relatively expensive, costing from somewhere
in the order of £50,000 upwards depending on sophistication.
However, as against this if you assume for argument's sake that
it works only five times faster than an output line printer the
savings in terms of computer printout time and, therefore, cost
are considerable. These have to be calculated for each individual
application but the cost of equipment implies that it will be economic for in-house installation only where a large data bank is
present.

Fortunately, COM is available on a service basis. The BNB *Books
in English* has already been mentioned, but this is a special application. The National Westminster Bank, Marketing Services Department, has operated a COM service for some time. At the time of
writing their charges are based on a minimum of £15 per 2,000
frames, for original film, an additional charge of £2·50 being made
for Kalvar copies, plus 75p for the cassette. The equipment used
is a Datagraphix 4440, which operates fifteen times faster than
most impact printers. The advantages which they claim for their
service are :

1 The slowness of paper printout.
2 The expense of additional runs for numerous copies.
3 The high cost of storage of hard copy printout.

While there is no quarrel with the first and last of these, there are
other ways to avoid re-runs to obtain additional copies. Computer
printout can either be duplicated by xerography or printed out on
to small-offset plates in continuous form and the copies run off
from these on a small-offset duplicator.

H

Planning a COM system

I suggest that in planning a COM system full advantage should be taken of the flexibility of microforms to ensure that all those who need the information obtain it as and when they need it. In planning any system whether in-house or using a service house to do the actual COM filming and processing the first consideration will be savings in computer printout time. Theoretically if a computer costs £x to run for a given volume of data, the saving will be £14/15x during the time it is printing out on to COM, compared to impact printing. In practice one seldom attains the theoretical maximum but this still leaves a considerable amount of leeway. Further, since the microfilm generated takes up only 2 per cent of the space taken up by an equivalent volume of information in hard copy computer output form, this alone should be sufficient saving to counteract the cost of supplying readers to those who need to use the information. These two factors supply the leeway in making computer generated information more generally available at no additional cost; and in most applications at a saving when compared to impact printer output.

Earlier I referred to COM as a 'poor man's time-sharing' system. As with most systems, speed costs money and the delay implied by processing, duplicating and distributing the film must be taken into consideration. This can be anything from about 2 to 4 hours where an in-house installation is envisaged and up to a minimum of 24 hours where a service house is used. However, there are many applications where the immediacy of an on-line system is not necessary and in these COM offers the possibility of a number of people sharing the same computer generated information at the same time. In this sense it is a 'poor man's time-sharing' system.

As with microfilm which is originated from any other source COM film can, once the master copy has been obtained, be treated in any of the ways already described. It can be distributed either in roll form, or jacketed and distributed, as fiche, or it can be turned into PCMI ultra high density fiche. Further during this process the master film can be copied in parts so that users get only the information they require.

Planning an efficient COM application is not, therefore, merely a question of transferring impact printer output on to microfilm unless, of course, the application is merely archival.

Inside the computer room

Quite obviously COM is not suitable for printing out all computer processed data. Previously I have stressed the need for a permanent relationship between the originator and user of microform held information. This rules out applications where the documents prepared on the computer are for distribution to the general public, for instance, invoices or monthly statements. You will still need an impact printer to perform this kind of operation.

Theoretically you could print out on to microfilm and then produce hard copy on a xerographic enlarger/printer, such as the Copyflo. It would, however, require a very hard look at the overall cost to establish the economics of this kind of operation. On the one hand you have a considerable saving in computer time and on the other the processing and duplicating of the microfilm (for file copies) plus the printout, plus guillotining the printout into individual documents. It might just show a saving where all the factors are favourable. Nevertheless in the present state of the art most users will be thinking in terms of using the microforms rather than regenerating hard copy from them.

However, this is only one aspect. If advantage is to be taken of the opportunity to provide a wider distribution of information, it is probable that some re-programming will be necessary, in addition to that needed to convert the original hard copy printout to a format suitable for the COM equipment. In this context it is notable that some, but not all, COM equipment is capable of producing graphics, which would require a separate plotter if the graphic printout were to be in hard copy form.

Exactly how compatible a COM equipment is with a given digital computer output depends on a number of factors. Some are designed to operate off-line, while others can operate either on-line or off-line to a mainframe computer. If the experience of users of visual displays is any guide, full compatibility should be investigated before a firm order is placed.

From a system viewpoint

From a system viewpoint planning a COM system will naturally involve both the computer room and those who are to use the information. It may be a personal reaction but I suggest very strongly that if people are given too much information they will not use it. This is all too easy to do when you have a means of accessing computer generated information which is as fast as COM.

The temptation is to distribute information to all and sundry whether they have any use for it or not. This should be avoided otherwise a surfeit of information will defeat its own object. It is only human nature for a busy executive to ignore information as long as possible if it is buried in a mass of data which is irrelevant to his own area of activity.

This implies that the information as it appears on the 'master' microfilm will have to be split up into those sections which are of interest to individual executives. One way to do this is to copy on to fiche and distribute the fiche accordingly. Depending on the kind of information generated this could present problems, but for the reasons I have already stated I believe they are worth solving.

Another aspect of this part of the system is the form in which the information is presented. As previously mentioned, some COM equipment is capable of generating information in graphic form. It is claimed that marketing men think in this form, finding graphs and charts much easier to understand than tabulations. On the other hand this could involve considerable programming effort before alphanumeric information can be presented in graphic terms.

It almost goes without saying that those who need to access the information need to be provided with microfilm readers and that the type of reader chosen will depend on the microform chosen for the distributed information.

As with other microform systems the provision of reader/ printers is a ticklish question. While some people will undoubtedly need hard copies for detailed study, or for use where there are no reading facilities, the unnecessary taking of hard copy should be discouraged.

COM and microform publishing

The BNB *Books in English* is one application in the field of microform publishing which is already under way. The possibility of computer stored information being used to produce periodicals which are 'slanted' towards a small group of readers or even individual readers has been discussed. In the near future it seems certain that the two concepts will come together.

Some people have interpreted the computer generated publishing concept as the printing of books, or periodicals in hard copy form, varying the content to appeal to different groups of readers much in the same way that local papers have different editions for various localities within their overall geographical area. What is far more practical, and more likely to happen, is that publishers will assemble all the available information in a particular field of technology or academic study, store this in data form and abstract from it on request all the information within parameters defined by the subscribers to the service.

Such a service might, for example, in the academic field cover British history. The file would in this case be so arranged that information concerning a particular period, movement or historical character could be accessed. This would then be presented either as a bibliography or in full, depending on the original purpose of the information service.

Using the COM technique it would be reasonably economic to print out in this way either on to roll film or fiche. Similarly the COM technique could be used in the scientific field, for instance, to access computer held patent files or even catalogues of components and materials available from various sources.

The types of COM equipment available

At present all available COM equipment is of American origin. Specifications tend to be modified as manufacturers gain experience and new models are being added to the range both by manufacturers already in the field and by newcomers to this kind of equipment. Because of the complexity of the process and the extensive backing services required to bring a new installation to maturity it is more than ever essential to choose equipment from a manufacturer who is well established in this country and who has a good reputation for after sales service.

A COM equipment represents an investment of from about £50,000 to £75,000 and in addition at investment in programming and ancillary equipment which can vary from comparatively little to as much again, depending on the application. Depending on the model chosen a COM equipment will typically :

1 Accept output from 200, 556 or 800 c.p.i., 7 or 9 channel tapes; alternatively on-line operation is possible with some

models depending on compatibility with the mainframe computer in use; paper tape or punched card input are other possibilities.

2 Print out on to 16 mm or 35 mm roll film; microfiche formatting is possible with some equipment (usually as an option); retrieval mark coding under program control is possible on some models.

3 Typography may be varied; this is confined to one font on all but the more sophisticated models but the font can be selected from a range by the user; medium, italic and bold characters can usually be generated and on some models large characters which can be read unaided (usually used for fiche titling); graphics can be generated by some models; the electronics necessary for this are usually an additional extra to the basic price.

4 The output medium is normally silver halide (photographic) film formulated for microfilm applications but in one instance (3M–Electron Beam Recorder) dry silver is used.

5 Output speeds (i.e. the speed at which the information is transferred to film) range from 41,700 up to 120,000 c.p.s.

6 Page format is typically 64 lines of up to 132 characters per frame.

In one case, the Electron Beam Recorder, the characters are transferred directly to the microfilm without the intervention of a CRT display and lens system. Software is normally available insofar as it is necessary for the operation of the COM equipment. This may be of two kinds:

1 A package to convert the print tapes produced by most computer operating systems into a 'print on microfilm' mode.

2 A set of routines which can be called into the main print program to help in the formation of a tape suitable for COM.

Sub-COM systems

To distinguish them from COM equipment of the kind already described I will refer to microfilm systems which film an impact printer output as sub-COM systems. This is not intended as a derogatory title.

In the smaller information distribution systems sub-COM has many of the advantages of a COM system without the investment needed for a COM system. The equipment is simply a flow camera of the type previously described which is fitted with a continuous form feed. Several such cameras are available at a cost of between £11,000 and £2,000.

Since the printout is that of the normal impact printer no alteration is necessary to the computer program and such sub-COM systems are suitable for distributing the information from office computer and the smaller general purpose computer installations. As with other systems they have the advantage that once the 'master' film has been made this can be treated in any way which suits the distribution requirements.

Most users of this type of system will probably have the processing, duplication and, if necessary, reformatting on to fiche done by a bureau, as it is unlikely that the volume of information passing through the system would justify in-house equipment. The exception is where the system is one of several operated by the same organization.

The sub-COM system is also applicable to a number of specialist applications in the field of scientific and medical research. The basic requirement here is that of a computer generated trace in hard copy form, which in its original form takes up a lot of room and is awkward to handle when it is being compared. This is especially true of research programs which necessitate multiple traces on the same paper roll output.

CHAPTER 13

Saving computer time
and memory

In the previous chapter I discussed COM as a process but indications are that one of the main applications for microfilmed computer generated information will be to relieve the computer from having to access back files, in much the same way as micropublishing is used to relieve the publisher from the necessity of keeping stocks of back numbers.

The high cost of interrupting current work on a computer so as to access information stored in digital form is common knowledge among computer managers. Whether this cost is equally appreciated by general management is open to dispute. Much depends on which side of the fence you are on. If you are the managing director and you need a stock level figure, you need it now. Never mind if the computer room has to interrupt the invoicing program they are in the middle of running to re-run a part of the stock control program. Further, electronic storage, whether tape, disk or drum, is fairly expensive compared to microfilm. Correctly applied a system which stores your back files on microfilm releases computer backing store for more immediate tasks.

In any information system there is a good deal of information which remains static over comparatively long periods and if this can be stored on a less expensive medium than the data processing system considerable economies can be made, provided it can be retrieved readily. Hence retrieval systems, such as Miracode and the multiple fiche systems of which that sold by Image Systems is an example; this consists of a number of barcoded fiche suspended in an enclosed carousel with an electromechanical device to enable any required frame to be retrieved in seconds via a keyboard.

For example, in a booking system, generally thought of as an ideal on-line computer application where a centralized data bank is accessed via a visual display (CRT) terminal, much of the information remains static for months at a time. To take three typical booking systems: on an airline, or for that matter a railway, the timetable remains static for at least six months; in a packaged tour operation the details of transport and accommodation remain static for a 'season'; in a theatre booking situation the accomodation plan remains static for years at a time.

In all these situations all the agent needs to know is whether the facility asked for is still free. To display the static information at the point of access saves valuable line time. It can also save valuable computer storage, since it is only necessary to store the variable information and the code which addresses it if the static details a customer may need to know can be read off from another medium.

Further, when the agent is suggesting a number of alternatives it is not necessary to access the computer until a decision has been reached. In addition microfilm is one on which graphic information can be imaged and displayed, just as easily as alphanumeric, an asset with some types of information in this general field.

In most financial applications it is not necessary to distribute the microfilmed historical file among a large number of branches, and consequently few copies are required of the master microform. Using the majority of such applications, the branch telephones the head office (or computer centre) where the main file is held. The clerk then accesses the microfilm file for the required information which can usually be found in a couple of minutes while the enquirer is still on the telephone. Provision of a reader/printer enables this information to be confirmed by post.

This basic situation can be applied in a number of areas, wherever it is desirable to relieve the computer system of the necessity for storing large amounts of virtually static information. A computer is basically a means of processing information quickly, of comparing or amalgamating information from two or more sources; its function as a store should be seen as an adjunct to its primary function and not as a primary function in itself. This may be disputed by some but the fact remains that in order to access, i.e. get information from a computer, it is necessary to use expensive equipment which works at mechanical speeds. Mechanical speeds are lower than electronic speeds by a ratio of at least $15:1$. Therefore printout is expensive in terms of computer time. Why, therefore, access the computer a number of times to obtain static or back-file information when you can obtain it once and access it quickly on much less expensive equipment? Admittedly you can, and most people do, keep this information in hard copy form, but storing it is expensive and we all know how long it takes to locate a particular item from a file of any size printed on standard computer stationery.

Additionally there are situations where a computer file can be relieved from holding static information, other than in the booking systems already mentioned. In most systems these blocks of information are accessed by reference to a code which is protected, so that it is only necessary to hold the information which is to be manipulated on the computer. For example, in a trading situation a code will represent a stock item. This stock item will not change, or put another way, if it is changed in any way it will be given a new coding. To hold a complete description on the computer would be extremely expensive and usually where a description is held this consists of at most three or four words. It is, however, feasible to hold a complete description with illustration on microfilm and to link this to the computer held information, using one of the automatic retrieval systems.

Such extensions of the information which can be held on a data processing file are becoming increasingly necessary if only because the amount of information available is growing at a rate beyond the capacity of the human memory. Earlier I discussed the BMC Service Department parts lists which run to 72 volumes. But today where retailing is carried on in ever larger units can a buyer honestly say that he can remember all he needs to know about the types or merchandise for which he is responsible? Or that all he needs to know can be economically filed in data form? Admittedly pricing information needs to be stored in data form as this can be quantified, compared and manipulated. Design information cannot be quantified except to a limited extent. Yet in many areas of merchandising the volume of design information is such that to hold this in hard copy form makes comparison virtually impossible and search a time consuming task, even where it is well indexed.

I do not suggest that microform files are the answer to every type of design information problem but I do suggest that it is well worthwhile looking at applications outside the servicing area whenever the volume of information available is becoming too large to handle conveniently.

Over the past decade we have taken to thinking that electronic data processing holds the key to solving every problem. We have almost come to believe that an electronic solution to a problem must be superior to any other solution. We are now beginning to realize that data processing is exactly what the term implies—a means of manipulating data, of performing mathematical manipulations, whether these relate to accountancy or technical development. As a duplicator a computer is the most expensive machine available; as a filing cabinet it is even more so.

While data processing is the heart of an information system it is not itself a complete system. In economics we used to learn that goods have no value until they reach the consumer. In the same way information is valueless until it reaches the person who needs to act on it. It is distributing the computer generated information that microfilm can offer economical solutions to many problems. In doing so it can extend the range of computer generated information in two ways:

1 By making information available to a larger number of people in the form that it 'comes off' the computer.
2 By supplementing computer generated information by information from other sources.

In this latter application it can relieve the computer from holding information which is either static or cannot be quantified and is, therefore, irrelevant to the computer program.

CHAPTER 14

Computer controlled microfiles

Microfilm records can be said to be under computer control when the index is on a computer. From this it is only a short step to integrate the microfile into the computer program. This is done by connecting the indexing program to the film transport on the reader. The film can then be motored automatically to bring the frame chosen by the computer program to the viewing position. This can be done either when using roll film or fiche. In the latter case any frame from a number of fiches can be chosen by mounting the fiches in a carousel or other multiple microform rack provided with the necessary selector mechanism. Indexing is usually based on a binary code which is either optically or magnetically encoded depending on the stage at which the interface comes between electronic and optical/mechanical operation.

From the user's viewpoint the advantage of computer controlled microfiles is that they enable static or slow moving information to be stored on film rather than electronically in the more expensive computer backing stores. Whenever you find an application where the file comprises a large volume of near static information and a comparatively small amount of fast moving information there is a potential for dumping the static information on to microfilm, while still retaining an active link with the computer held part of the file. This is particularly so where each block of frequently updated information needs to be backed by a large amount of related static information.

There are a number of instances of this situation in different industries. For example, it would be feasible to design a mixed microfilm/data processing system to hold the specification of a product, in which the basic specification is held on microfilm and any subsequent modifications are held on the data file. A microfilm terminal would then print out any relevant modifications added after the original specification had been written when a particular page or block of pages was accessed. This would be cheaper than keeping the entire specification in data form. Equipment for this type of mixed system is already available.

Again there are a number of directories in which only a small proportion of the information changes over a given period and it would be feasible to devise a system whereby a directory was held on microfilm. Any changes after the date on which it was last compiled would then be held on the computer and printed out when the relevant page was accessed. The directory could then be updated on microfilm at convenient intervals and the computer update re-started from that date.

Under some circumstances booking systems can be devised along the same lines. For example, air schedules usually change only twice a year and at least one airline in the United States uses an on-line computer booking system for their own flights supplemented by a microfile of the schedules of all connecting flights operated by other airlines.

Technically there would be no difficulty in setting up a similar system for the packaged tour industry. Using such a system a travel agent would hold all the tour details on microfilm and access a central computer booking system which would then only need to hold the tour codes plus availability status information. Whether the travel industry would be prepared to co-operate to this extent is another question. However, the basic characteristics are present and the alternative, an on-line system using visual (CRT type) terminals, would be far more expensive to operate.

These are just a few of the situations where an active system could usefully combine microfilm under computer control. Other examples occur in the financial and insurance world. Much depends on how you need to sort information when considering whether it can be stored on microfilm. For instance, if the application is such that the categories into which you need to sort the information can be contained within the code which is also used to identify its position in the microfile, you are indeed fortunate. On the other hand, if sorting requires the storage of a considerable amount of information you may find that the creation of a separate microfile accessed by the computer is uneconomic, simply because you have to file this information in data form anyway.

However, if this information is static there are other ways around this problem. A French company has in recent years been exhibiting a microfile system which incorporates a binary bar code, relying on optical coincidence. This enables a large amount of information to be stored on each fiche which can be sorted in almost as many ways as if the information had been on an electronic data storage medium. One application suggested for this system is in police work as, in addition to the encoded information, the fiche can store a photograph and/or a certain amount of text.

Microfilm/computers and training

One reason which is holding up computer-aided instruction is the expense of installing and maintaining the electronic data processing equipment. Unaided microfilm can simply guide the student from one frame to another, but the addition of even a simple mini-computer enables the learning to be programmed. How sophisticated the teaching program can be depends on the ingenuity of the programmer as well as on the power of the computer.

The simplest system gives a plain 'yes' or 'no' or, if you prefer, 'right' or 'wrong' choice together with a program lock to prevent the student going forward until each step has been completed correctly. However, there is no reason why search programs cannot be devised whereby a student is required to relate several pages of information.

Another probability where a keyboard terminal is incorporated in the microfilm reader is to set up a program whereby the question is posed on the microfilm and the student enters the answers on the keyboard. This enables a program to be set up whereby the student is required to study several frames of film and to answer one or more questions asked on the last frame of the sequence.

Admittedly the equipment required for this type of program is fairly expensive and the main advantage lies, when compared to a visual terminal, in the flexibility possible in the way the information is presented on the microfilm. It can, for example, include photographs, line drawings, graphs, charts and artwork of all kinds which would be difficult if not impossible to store in digital form.

Exactly what the function of computers in learning will be, will probably not be fully appreciated for at least another generation. In the meantime those concerned in the educational field can but experiment with what is available to them. If assertions heard from time to time that children like computer guided instruction because the computer is not subject to the emotional failings of even the most patient human teacher, then, for some purposes, computer-linked microfilm would seem to have a bright future. It combines the patience of computer-based instruction with a medium which is far less expensive to prepare than a completely digital data instruction program.

It is, for example, conceivable that a program can be devised which guides the student progressively from one frame to another under computer control which will be adaptable to a number of different subjects. This would give the lecturer or teacher the possibility of preparing his own material, without necessarily having any knowledge of programming.

All this must be largely speculative since it has not, so far as I am aware, yet been tried. But I suggest that if trends which are already present in both computing and microfilm the following will be possible within the present decade on a limited scale and thereafter more widely:

1 Computing is gradually developing on the lines of a utility, that is to say in the future you will be able to 'plug-in' to computing facilities as you now plug-in to the electricity mains.

2 Any organization which needs computing facilities will eventually be able to have them 'on tap'. Quite obviously, at least in the near future these facilities will not be cheap.

3 It will, therefore, only be practical to use computing utilities for education and training in the most economical way which will achieve its purpose. This will apply both in general education and to training within industry and public administration.

4 While there are some subjects which by their nature require direct dialogue between computer and student, this does not apply in all cases. In fact for a number of subjects and in certain levels of learning the combination already described has advantages over the dialogue between man and machine possible even with visual terminals.

5 This indicates a place for computer/microfilm instruction somewhere between full on-line computer instruction and television instruction. This latter allowing no response between student and source of instruction leaves the student virtually on his own. It is, of course, possible that immediate response can be built into a television-type transmission system,

especially where it is a cable, i.e. closed circuit transmission, but for all practical purposes you are then back with a video terminal system to which live or video tape transmission has been added.

6 Used exclusively such a system would result in extremely rigid education reducing the local teacher to the role of disciplinarian.

7 In view of the computing utility mentioned above (1), a computer-controlled microfilm program which can use material prepared by any teacher or lecturer would restore the teacher to his full function. It would, for example, enable a national instruction aid program put out over television to be supplemented by a local program, prepared according to the abilities and interests of a group of students with whom the lecturer is personally acquainted. It is true that such a program can equally well be devised without the computer control, leaving the students to find their own way through the information available. While this is acceptable and even desirable at the higher educational levels, at the primary to middle level (and corresponding early stages of industrial training) the brighter pupils should, if the program is well written, receive the aid they need from the machine, leaving the instructor free to concentrate on the 'lame ducks'. This, at least, is a potentially viable concept which does not turn all teaching into impersonal instruction.

CHAPTER 15

Microfilm in the drawing office

In the relevant British Standard BS 4210 it is assumed that drawing offices use the A size originals and although many drawing offices have now gone over to the use of these paper sizes the practice is not yet universal.

While the use of A size originals is not essential to a microfilm storage and retrieval system it is desirable if only because the frame size has the same ratio as the A size originals. However, many companies, especially in the USA, are successfully using microfilm in their drawing offices where originals are drawn on 'traditional' paper sizes. The main argument for adopting A size originals is the general one that metric units are now used by the engineering industries in this country rather than in any insuperable difficulty in microfilming other sized originals.

For all those industries where A sizes are adequate for the purpose of containing the information it is necessary to put down in the drawing office, the normal storage medium is 35 mm aperture cards, each drawing being mounted in a separate card.

In this A size system all originals are drawn on five sizes from Ao down to A4, the latter size also being used for specifications. This system also has the advantage that all originals can be filmed at one of five standard reductions to fill a 35 mm frame which makes the operation of filming the drawings considerably more efficient than where the old imperial sizes are in use.

The management viewpoint

From the management viewpoint changing to A size originals and to a microfilm storage and retrieval system has several advantages but also needs to be carefully planned:

1 During the initial changeover period there will be a large 'back-file' of drawings to be microfilmed. These should be microfilmed by a bureau so that the changeover period is kept to a minimum even if in-house camera facilities are planned from the start.

2 Draughtsmen will need to be trained in the requirements of preparing drawings for microfilming. These requirements are defined in BS 308. They are, in fact, common sense and can be assimilated by an average draughtsman quite quickly.

3 In a unionized drawing office it may also be necessary to assure the union representatives that the changeover will not affect their members' job status.

The following benefits can be expected:

1 Lower print costs. Where hard copies are kept to half the linear size of the original the print cost per copy is actually less than when full-sized prints are made on diazo equipment.

2 Fewer copies will be needed. In many cases draughtsmen will merely need to consult an aperture card in a reader. Some precautions should, however, be taken against draughtsmen building up their own personal files as this will largely defeat the object of microfilming.

3 There should also be some increase in productivity. Some of the time wasting practices such as close-hatching and fancy lettering cannot be tolerated by microfilm systems; they do not reproduce well. It might be supposed that these practices had died out some time ago but this is apparently not true of all industries.

4 There will, of course, be a large saving in space.

Microfilming old drawings

It is usually in the changeover period that troubles occur, as it is frequently necessary to microfilm originals which do not meet the necessary standards of clarity, i.e. include words and figures with thick and fine strokes, fine lines or a combination of both. In such cases it is necessary to microfilm at a lesser reduction than would be used for a drawing drawn for microfilming, even if, in the case of large drawings, this involves microfilming them in sections. Alternatively, where there is a large proportion of old drawings which will not stand the necessary reduction the 'back-file' can be kept on 70 or 105 mm film. These are usually unmounted and kept in translucent envelopes to protect them against dust and scratches.

Usually you will find that the large format is necessary only for some assembly drawings and although it will be a little more ex-

pensive in the first place it makes a neater arrangement to have all
the large drawings up to a given date on the larger film, rather
than to have a mixed file. In any case it is desirable to have the
older drawings microfilmed by a bureau because they represent
a 'backlog' which will not recur. Further, they are the ones which
will need the most experienced camera operator if the best image
is to be obtained from them.

Destroying originals

One problem with microfilm is updating original drawings to
incorporate modifications. Some manufacturers claim that you can
obtain up to twelve or more generations using their materials and
equipment. By this they mean that you can print out a full-sized
copy from the first microfilmed original, modify it and re-microfilm
the modified drawing—and repeat this twelve times or more.
BS 4210 states that 'the quality of each microfilm image shall be
such that every line and character of the original is recorded on
the microfilm, with sufficient contrast and clarity to be reproducible
up to and including the third generation.'

This may be a little conservative but with records which are
as costly to originate as engineering drawings it is better to be
safe than sorry. In fact it is better not to destroy the originals
at least until the product is out of production. It is not the originals
which take up space but the size-for-size copies.

The number of times a drawing can be modified using a microform as the original depends on several factors not the least of
which is the suitability of the original drawing and the care with
which the draughtsman makes the modifications. Generally speaking the bolder and more 'open' the original the more it will stand
modification.

Technical considerations: originals

Drawings which are made on traditional materials can usually
be microfilmed satisfactorily but the modern film materials are
particularly suitable for this purpose. Whatever instrument is
used to mark the original this should give a good contrast and
the drawing should be done in accordance with BS 308 supple-
ment 1. That apart, there is little restriction imposed on drawing
practice by the subsequent requirements of a microfilm storage
and retrieval system.

Storing and retrieving engineering drawings

As with any other microfilm system it is usual to use the original roll film as the archival copy. This is then duplicated on to either diazo or silver halide film and the duplicates are mounted in the aperture cards. Except in the larger drawing offices it is usually necessary to keep only one file which can be arranged in exactly the same way as a normal hard copy file and indexed accordingly. The index itself can be either in hard copy form or itself on microfilm.

However, it is possible to be too economical and if file copies of the drawings are not readily available there will be an increased demand for prints made from the microform copies. For the same reason an adequate number of readers should be provided for the use of the draughtsmen.

The facility provided by the aperture cards for sorting drawings automatically is used by comparatively few drawing offices. Whether this is because suitable card sorting equipment is not readily available or for some other reason is hard to say. However, if this facility is used it is better to punch a dummy set of cards as sorting equipment can be quite rough on apertures. Once located the required card is then pulled from the file manually in the usual way.

Engineering drawings can be retrieved in hard copy using one of several processes; electrostatic is the one most commonly used today because of its convenience. Equipment to print out from aperture cards right up to A1 is available. There is also equipment available for printing directly out on to diazo materials or the micro-image can be printed on to a translucent material which is then duplicated on a normal diazo machine. This method is the most economical where multiple drawings have to be supplied to other drawing offices which do not have microfilm facilities.

Copies for other departments

These days there are a number of departments loosely associated with the drawing office which use engineering drawings—technical servicing, specification writing, training, etc.—even marketing in some industries. If the number of hard copies demanded is not to become excessive these departments will also need readers and duplicate microforms.

There is also no reason why the engineering shops should not themselves use microforms. After all, the only reason for needing a size-for-size drawing is when it is used as a template. Do your inspectors measure from the drawing? I doubt it. In most cases they 'take off' the stated dimensions, tolerances and other relevant 'alphanumeric' information.

The limitations of microfilm

At the present time microfilm techniques are not sufficiently developed to be used with some of the other techniques used in the drawing office.

Scissors drafting

Scissors drafting quite obviously presents a number of difficulties in the reproduction of drawings. Where an old drawing is taken as the starting point and an area cut out of this which is then modified to produce the new drawing there will be some difficulty in reproducing the final drawing using a medium such as microfilm which requires a clear boldly drawn original. This difficulty is accumulative in the sense that after a while successive modifications made in this way mean that eventually none of the original drawing will be left, while various parts of it will be of different generations. Accepting that there is some deterioration in quality with each generation sooner or later the drawings become unreadable and it is necessary to trace them by hand.

Photodrafting

Photodrafting is another technique which cannot be used where microfilm is the reprographic medium. This is the technique whereby a photograph is taken of the original situation and modifications are added in the form of a 'sketch'. To reproduce such drawings, which contain continuous tone, it is necessary to screen the photograph. The screened photograph is then reproduced on to sensitized drawing film and the modifications indicated by normal drawing office techniques in the border area surrounding the photograph. The whole is then reproduced using any of the normal size-for-size techniques designed for drawing office use.

Apart from these two techniques the only other restrictions imposed by microfilming are those concerning the standards of draughtsmanship which have already been discussed.

CHAPTER 16

The overall reprographic service concept

In most organizations there is still the concept of the 'duplicating unit' or the 'internal print department' as an entity, and various copying or duplicating facilities provided throughout the organization as something quite else. Phrases such as 'we have our own copier' or 'we use stencil but I think they have offset in the print unit' are all too frequently heard. When contact is finally made with the 'print unit' or 'internal print department' it is often found that the work is orientated towards the department which it was founded to serve. The print manager is often an 'offset man' with little interest in other copying or duplicating processes. He is quite genuinely convinced that if only management would make more use of his printroom, stationery bills would be cut and everything would run more smoothly. He is, of course, right in one sense but his thinking is confined within his printroom. He is thinking of himself as a producer of small-offset print and not as a provider of visible communications. The need is there and it is probably not his fault that he is not meeting it. Much depends on his original brief; on how his superiors see his function in the management team.

It is still too frequently overlooked that the need is for visible hard copy communication. How this is achieved is of secondary importance. I hope that it has become abundantly clear by now that there is no one answer to the copying and duplicating problem which is valid in every case. To serve an organization of any size a mixture of processes is needed if every requirement for visible communication is to be met economically, yet all these need to be tied into one system. If the printroom manager is not willing to come out of his printroom and supervise the entire operation he will inevitably find that he has a manager over him who is willing to do so. Admittedly the concept of a communications manager can only come from the top and, although such a person exists in some organizations already, he is still a rarity.

I have already stressed several times the fact that information is useless until it reaches the person who is to act on it, even if this action is, in some cases, negative. Further, the speed at which this information can be distributed will affect the efficiency of the organization. This does not mean that all information must be distributed as quickly as possible regardless of cost; the problem is not as simple as that. If it were we could all change over to electronic data transmission immediately and feel the benefits.

In practice there are a number of situations in any organization where the speed of the information flow is regulated by the process which it serves. For example, if all the monthly statements are ready for mailing by the end of the first working day in the month no added benefit is apparent by having them ready within the first hour of that day since they will not arrive at their destination any quicker. There is no point in having all incoming orders processed and ready for the warehouse within say half an hour of their receipt in the morning mail, unless the warehouse is arranged in such a way that it is more efficient to deal with all the day's orders in one batch. Otherwise it is better to arrange for a steady flow. It all depends on the kind of product you are handling.

Whether these examples are valid for you is beside the point, which is that every organization has areas where, because of its nature, the work flows more slowly than in other areas. The job of the communications manager is to identify these areas and to arrange for the documentation flow accordingly. Whether telecommunication needs to be considered depends largely on the distance between the source of the information and the place where it is to be acted upon, and consequently on the size of the organization being served. In a large organization such as a government department or an international industrial group, therefore, the communication manager will need a knowledge of telecommunications as well as hard copy communication, which means virtually copying, duplicating and microfilm.

Today, telecommunications are a big subject in themselves and I do not propose to discuss them in any detail, but even in the largest organization there comes a stage where the electronic impulses which are the lifeblood of telecommunications have to be turned into hard copy readable by the human eye. In the largest organizations the overall communications manager may need an

assistant to supervise the hard copy stages of the complete communication chains, but in smaller organizations and ones where
the immediacy of telecommunications is of less benefit the communications manager may be concerned largely with hard copy
communication. It is this that I am concerned with at the present
time.

There will, of course, be some telecommunication in even the
smallest industrial, commercial or professional firm. We all use
the telephone and facsimile links are useful, to say nothing of the
public teleprinter service. However, in smaller organizations we
can still, in the present state of development, treat these as 'off-
line', that is, we can tolerate a transcription stage between the
output of the telecommunication link and the input to the system.
This is just a 'technical jargon' way of saying that, for example, if
we receive an order over the Telex it will not hold things up too
much if we transcribe it on to an order form rather than feed it
straight into a data bank or other electronic storage device.

To get back to practical considerations, let us say that at the
moment we have an internal print department, under a print
manager. Dotted about the company there are also various copying
and duplicating facilities which are loosely under the supervision
of the stationery buyer but in practice looked on as the sole
property of the department to which they are allocated. In addition,
there may be copying facilities which are common to all those
working in a building, the so-called centralized copying pool. Is
this an efficient way of dealing with the communications problem? In many ways it is rather like having an accounts department
which deals only with the larger accounts while each department
deals with its own accounts in its own way, only relying on the
accounts department when it cannot cope with the task.

Being only human the print manager naturally wants to expand
his own department and enhance his own status. He will, whether
he realizes it or not, be against the introduction of any system or
process which does not require the kind of print he can produce.
Departmental managers, on the other hand, will tend to demand
individual copying or duplicating services. This again is only
natural. We all like to think that our own work is important
enough to require the best tools and in the eyes of the departmental manager the best copying tools are those which only his
department uses, since his workflow cannot then be held up because

someone else has claimed priority over a communal copying facility. Not all companies let these things happen but the tendency is there because what is basically the same function is divided in the management function. Look on copying, duplicating and micro-film as different forms of hard copy communication and it becomes obvious that they should all be under the same supervision. Although each copying or duplicating task may be a part of a different paperwork system they all perform the same function to communicate visibly in a permanent or semi-permanent form.

It can, of course, be argued that copying incoming correspon-dence on an office copier is so far removed from, say, the duplicat-ing of service manuals by small-offset that the two have little in common, but they are both ways of communicating essential information to those who need to act upon it. Otherwise why make the copies?

Now let us look at copying, duplication and microfilm as a service, not to one or more parts of an organization but to the whole.

We have all read about integrated management services based on a computer. This involves the use of a central processor and data store into which information is fed and retrieved. According to the requirements of a particular situation this data may be fed directly into the computer or may be captured at some place remote from the computer and fed into it either directly through a data terminal or indirectly by another machine which punches a tape or punched cards. In the latter case these are conveyed physically to the computer room. In computer language the data may be captured 'on-line' or 'off-line'. Data is merely in-formation in machine form. What we are concerned with is information in hard copy form. We can now forget the computer, as such. Think of the printroom, internal print department or dupli-cating unit, whatever you like to call it, as the computer room with its data bank, i.e. with the facilities to provide preprinted forms and deal with any duplicating needed in comparatively long runs, and the processing of microfilm.

Now think of the various departments in the organization as the peripherals. (See Figure 8.) Their hard copy requirements will vary both in volume and immediacy. Some will need immediate 'on-line' access to the central copying unit, others will not. Some will need their copying facilities integrated with the machines which

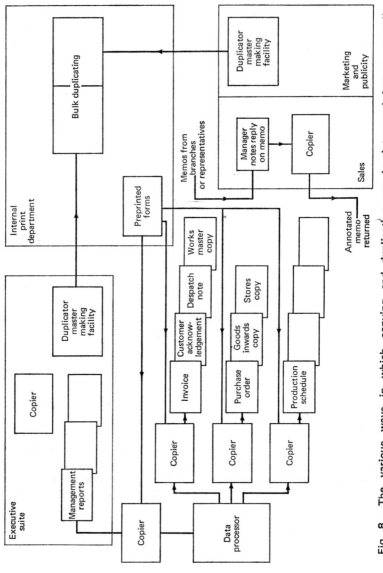

Fig. 8. The various ways in which copying and duplicating can be integrated into an overall documentation system.

produce the master to be copied, but in all cases if the copying facility is to be used economically they will need it to be integrated with the paperwork system which they are operating. It is this which is the task of the communications manager so far as hard copies are concerned.

How this is done on a human level will depend on management philosophy. Certainly the communications manager will have to be persuasive, as inside the company he is selling a service. Freed from the confines of the 'internal print department' he can offer the most appropriate answer to each copying problem, regardless of the process. He can, in fact, offer any combination of the following:

1 In a department where figures are being manipulated a facility for manipulating words which is integrated with the machinery used to manipulate the figures.

2 In a department where words are the main stock in trade both a self-contained facility for copying words and an input to the central copying unit.

3 Any mixture of the above which will solve a particular copying problem.

Now let us see how this works out in practice. First I will detail an application in the sales documentation routine, to which reference has been made in previous chapters. It is, after all, the routine which is common to all commercial organizations.

This is an application where data processing and word processing interlock, copying taking over from calculating once the mathematical process is finished. Starting with the receipt of a customer's order, the mathematics may be carried out on an invoicing or other direct entry computer or they may be done 'in the head' with the aid of a calculator. If a direct entry computer or other keyboard accounting machine is used it will produce a document which can be used as the master for the copying or duplicating steps which follow. Means, which have already been discussed, are available by which unwanted information can be deleted from some documents and other information added. Most of the documents produced on the first run go to the customer, representative or branch manager as records that the order has been correctly received but one will be the works order. This document may itself be the master for a further set of documents needed for production or purchasing, and these too can be raised reprographically. However, on one of these we return to a mathematical step where the order is priced and this can be fed back to complete the customer's invoice.

Where such systems combine both data and word processing they can be considered as 'closed loops'. The reprographic equipment needs to be sited adjacent to the data processing equipment so that the process can be continued without a break, although, if preprinted forms are required these can be provided by the central duplicating unit and processed on site.

Regardless of the kind of equipment used to provide the data the communications manager has several options as to the process he will use for the reprographic steps in the paperwork chain:

1 He can use a copier/duplicator of the xerographic type.
2 He can use small-offset.
3 He can use spirit duplication.
4 He can use any combination of the above three processes.
5 Even here his choices are not ended, since he has several options as to how the masters will be made if he uses small-offset or spirit, direct or indirect and, if indirect, by which process.
6 He can use microfilm as the storage and retrieval medium.

Certainly all these options are not open for every piece of data producing equipment. Nevertheless, with every piece of data producing machinery, known to me now, from a calculator to a full-scale EDP system there is a choice. Provided his choice is acceptable to the chief accountant, or whoever is in charge of the invoicing procedure, and to the works manager, or whoever is in charge of the stores and works procurement and production control, the choice of process may be governed by factors arising in other parts of the organization. However, as readers will have noted, this sales documentation involves at least two and probably three senior executives and it would be one of the duties of the communications manager to co-ordinate the reprographic methods used by each. It can be said that this is the province of the organization and methods manager and certainly where one is employed the communications manager would work closely with him, but the organization and methods manager usually has a large number of tasks to perform and once the system is working would no longer be able to take a day-to-day interest in it.

We will now look at the reprographic requirements of a sales office which includes sales promotion among its activities. Such an office usually deals with enquiries from customers and the firm's own representatives, with sales statistics and with promotional material, although in a larger organization there may be separate sub-units.

A way of dealing with estimates reprographically using a printing calculator has already been suggested. Nevertheless this involves close liaison with the estimating department attached to the factory, another instance of two reprographic procedures needing to be co-ordinated. Is it cheaper for the sales estimator to send the production estimator a multipart interleaved carbon set or a master on which the factory prices can be filled in? It depends on the length and kind of estimate required, and on whether machinery is used to produce either or both estimates.

Again, if modern electronic accounting is used, the sales statistics will be a by-product of the invoicing procedure. Either the raw statistics can be sent to the appropriate section of the sales department which will then be responsible for distributing them as required or the complete job can be done as a by-product of invoicing. For example, the statistical output of an invoicing computer can be copied in whole and/or in part and the statistics produced in this way sent directly to those whom they concern. Again this is a choice involving two heads of departments. It also depends on the kind of reprographic equipment in use and its capacity to handle this kind of work in conjunction with its main task.

We now come to sales promotion. Here the requirement is likely to be for bulk copies. Notifications of price changes, direct mail shots, etc. all have to be distributed to comparatively large numbers of people. For this purpose the sales department can be considered as an 'input peripheral' for the central copying unit or internal print department. The problem here is where the responsibility of the sales promotion people should end and where it should be taken up by the internal print department. Should, for example, the preparation of direct image masters be the responsibility of the sales promotion unit or should the internal print department prepare these from copy submitted? It is more convenient to have plate-making facilities actually in the sales promotion department? This is unlikely but it is possible to envisage circumstances where it would pay off. If the print requirement of the sales promotion department is under the supervision of a communications manager who is responsible for all reprographic services these questions can be solved impartially. The department can be provided with suitable equipment to produce urgently required work if necessary, but the bulk of the work would be handled by the internal print department which would allocate priorities accordingly.

A similar analysis can be made of the reprographic requirements of every other aspect of a firm or an administrative unit in the public sector. In nearly every case it will be found that these requirements are dependent on those of another department, if only in the design of a form. In fact the purpose of copying implies that there will be interdependence since what one department originates others have to act on, and the way in which it is originated will have a bearing on the ease with which this action can be taken.

Hence there is a need for a reprographic manager. His duty is to co-ordinate the reprographic requirements of the complete organization and liaise with the organization and methods department to ensure that the communication requirements of each system they install are met in the most economical way. This is a far cry from the traditional concept of a reprographic or internal print department manager whose responsibility starts and ends with the copying department or printroom.

CHAPTER 17

The central reprographic unit or internal print department

It is usual to call the place where the high volume duplicating is done the 'internal print department', but 'central reprographic unit' is perhaps a more appropriate name as the words 'internal print department' imply that it is there to perform the same function as a commercial printer. In most cases this is simply not true. The primary function of internal printing or duplicating facilities is to fill the gap between what can be done in the office itself and the long-run high-quality work at which the commercial printer excels. The extent of this gap depends as much on what commercial services are available as on the requirements of the firm using these facilities.

Invariably, at any rate in this country, commercial printers turn out a reasonable standard of work except perhaps when the job is being done at a cut price; in that case you cannot complain if you get what you pay for—you often get more. What does vary considerably is the kind of services available in a particular district and in some instances the promptness of deliveries. In these circumstances a central reprographic unit which is appropriate to one company may be unnecessary for another. There is no hard and fast rule.

Generally speaking the shorter the run the more closely the duplicating work is tied to other processes on a time basis, and the more likely it is that it will pay to do it internally even if the cost per copy is higher than a commercial service would charge for the same work. Each case must be based on its own merits and in many a final decision on purely economic grounds is not easy to make. This may sound a peculiar statement to make, but although it is easy enough to compare the actual cost of getting the job done, how much is having to wait an extra 24 hours worth? It depends on what could have happened during that time. It also depends on whether better planning could have allowed a job done externally to be delivered at the same time as if it was done on the firm's own premises. There are so many of these ifs and buts. In setting up or redesigning an existing central reprographic unit they all amount to this:

1 What are the organization's requirements for duplicated and printed work, excluding work such as systems duplicating and short-run copying which can only be done conveniently in the office where it originates?
2 What urgency is there for this work?
3 In what lengths of run is it required, i.e. the number of copies from each original?
4 Any special requirements such as the need to reproduce photographs, the use of several colours on the same page, special size or weight of paper or card, etc., etc.

An examination of these factors will enable the right kinds of commercial print services to be sought and, if found, their reliability checked.

The difference between the services available and those required will give the 'reprographic gap' which has to be filled by the central reprographic unit. This will, of course, vary accordingly to the circumstances found in a particular firm.

One reason for referring to this service as the central reprographic unit rather than the internal print department is that the processes which can most economically fill this gap will vary accordingly to the kind of duplicating the unit has to undertake. Of those available small-offset is the most flexible but if this flexibility is not necessary alternatives may be considered:

1 Xerography, using the larger copier/duplicators which are capable of an output of around 3600 copies an hour, has the following advantages, or so its manufacturers claim:
a The equipment is simple enough to be operated by junior staff, only a 'key operator' requiring any special training which can be undertaken in a day or so.
b The masters are on plain paper and so are easier to prepare and correct than masters needed for other duplicating processes.
c When the equipment is used for duplicating rather than copying the overall cost is no more than about 1p a copy.

This cost is about in line with stencil duplication and a little above small-offset. However, to keep the cost down to this level it is necessary to do nearly all work on the lowest meter charge:

d On the other hand, all originals copy black and the quality of copies, while good, is not quite up to the best small-offset work, although better than much work produced by this process. The originals should be limited to line work as photography colour wash, etc. does not copy well.

2 Stencil duplication, although rather more expensive than small-offset to run has the advantage that it requires less skill and consequently staffing is less of a problem. Today when there are several indirect methods of making masters the process is more flexible than formerly. If the heat transfer method of making masters lives up to its early promise this will give flexibility, so far as the kinds of originals which can be copied are concerned, when the installation of an electronic stencil cutter would not be justified by the volume of work required.

However, by definition, a central reprographic unit is primarily a backing store for the reprographic services throughout the organization and where the volume justifies the outlay small-offset will probably be required because of its flexibility. It is largely a question of whether the equipment should be all small-offset or small-offset plus something else.

It is possible to make substantial savings by printing work internally but these savings must come from a combination of at least three factors:

1 The commercial printer is entitled to his profit and you can save this mark-up provided your internal costs, including over-heads, are no greater than his.

2 Using 'short cut' methods such as typing directly on to paper plates where the outside printer would use and charge for metal plates made photographically.

3 Making only the exact number of copies required, especially when doing short-run work. When buying print outside it is often necessary to order a minimum number of copies and in any case tempting to overorder.

K

I have already discussed the concept of the central reprographic unit as a 'backing' service for general copying requirements. To set up a unit of this kind it is necessary to meet certain requirements:

1 The unit should be situated where there is good natural light— a north light is best. If this is not possible artificial light which approximates to daylight should be provided.

2 General conditions of temperature and humidity should be reasonably consistent. Although desirable, air conditioning is not necessary but considerable difficulty will be avoided if the paper is stored under similar conditions to those in the duplicating room. Paper is a 'living' absorbent material which tends to shrink, stretch and to some extent change its characteristics under different climatic conditions.

3 Conditions in the printroom should be such that it can be kept clean easily and a reasonable working space provided.

4 Efficient working is aided by establishing a smooth workflow even in the smallest unit. Raw materials in the form of plates and paper should come in at one end and finished work emerge at the other.

In practice the normal office or factory environment will serve well enough but it cannot be too strongly emphasized that cleanliness and neatness are very desirable qualities in a room where duplicating takes place. Much of the success of a duplicating department depends on these qualities, particularly insofar as they encourage careful workmanship.

The need for care already emphasized in regard to small-offset is equally necessary when any other duplicating process is in use if waste is to be avoided and the best possible quality copies obtained from the equipment.

This need to take care runs right through duplicating from the preparation of the master to storing the final copies. Although there is nothing too difficult about any of the duplicating processes, even small-offset, care taken at every stage is worth more than high speed. With duplicating it is indeed a question of 'more haste less speed'.

I have emphasized this because it gives a good idea of the kind of staff to employ in a central reprographic unit. Steady workers willing to take everything in their stride but still adventurous enough to want to tackle a difficult job or master a piece of new equipment are ideal. This care needs to start with the preparation of the final copy of the original. When typing from immaculate copy it is much easier to avoid mistakes and, of course, this is essential if the original is to be copied photographically or mechanically. It is in fact mistakes which mar the appearance of more stencil duplicated work than any other single factor and even xerographic copies can only be as neat as the original.

The range of equipment suitable for use in a central reprographic unit is extremely wide and it goes without saying that a choice should be made on purely practical grounds. The final choice of duplicating equipment will be influenced by the following factors:

1 The largest size sheet of paper you wish to duplicate on to.

a Generally speaking, the larger the sheet handled the more expensive the equipment will be and the heavier it will be to operate. A stack of sheets 17½ by 22 in is about the largest size handled by the small-offset process and is heavy enough to require a man to handle it.

b On the other hand, if your operator has sufficient expertise considerable savings can be effected by duplicating small format jobs two or four copies to a sheet. In print language, two or four to face.

2 Whether you wish to use two or more colours on a sheet and/or want your print to be in exact registration. For this a machine with suction feed and other equipment to ensure exact alignment 'hairline register' will be required. This costs considerably more than a machine with friction feed of the same format. You can print two or more colours on the latter type provided you leave white space between the colours sufficient to cover any deficiencies in registration.

3 The average number of copies you require from each orginal; particularly when using small-offset for short runs of, say, under 250 copies the 'makeready' time is important, even when it is measured in minutes.

4 When using small-offset, if you wish masters can be obtained in the following ways:

a A number of trade platemakers exist throughout the country. Usually they will only make metal plates either from originals supplied or from 'roughs', i.e. they will supply the artwork at an additional charge. For prestige work or for those using small-offset for the first time their services are recommended. The only exception is when making short-run direct-image paper plates.

b The photodirect method enables plates to be made in about two minutes, and incorporates facilities for reducing or enlarging the original. Other, less expensive, versions of this method are available for making plates from same size originals.

c In addition to xerographic platemaking equipment there is another kind of electrophotographic equipment which makes plates in a few minutes to a very high standard. Also paper plates can be made by this method.

d Auto-reversal film, which is slow enough not to spoil in subdued daylight can be used. A special exposure unit is required first to expose the film to the original and then to expose the 'negative' to the plate. No darkroom is necessary.

e There are a number of simple process cameras on the market which can be used either with conventional materials or with projection speed diffusion transfer materials.

f For 'size-for-size' copying a diffusion transfer copier can be used.

g Some electrostatic copiers can be used to make paper or 'plastic' plates, depending on availability of suitable materials.

The price of equipment made for platemaking as opposed to machines which are primarily office copiers, varies in a ratio of about 1 to 10, the cheapest being a diffusion transfer copier costing from about £100 to £200. The most expensive is either an extremely accurate process camera capable of making plates to fully professional standards, although photodirect and other rapid platemaking apparatus is in the same price category. It is only fair to add, however, that as with all photographic apparatus the cheapest is often the easiest to use and will give reliable results provided the copying is straightforward. It is if you want extreme speed, or the ability to do everything a commercial platemaker can do, then you have to pay for it.

Apart from the characteristics already discussed, the advantages of small-offset in a central reprographic unit lie both in the variety of materials which can be printed and in the range of direct and indirect imaging plates available. The latter enable runs of from as few as 50 copies, some claim less, up to virtually unlimited runs to be made economically on a non-system machine. The latter can be made over a period in a number of separate short runs. The former enables most papers to be used over a wide variety of weights. On the more expensive machines anything from airmail to three or four sheet card can be handled, although very light or very heavy papers require some experience of the process. Medium weight papers such as a normal bond are probably the easiest to handle but rough surfaced papers print well and even glossy art papers which have been formulated for offset work can be printed. In addition, all kinds of special papers and even some types of plastic sheet can be printed, but again a careful and experienced operator is needed. (See Figures 9, 10, 11 and 12.)

On the best kind of small-offset machines it is not difficult to print process colour work (i.e. where four screened colours are superimposed to give a full colour effect) but, although this is frequently done by commercial printers, few internal print departments attempt this. By the same token when used by an experienced operator all kinds of speciality inks can be used. I have mentioned these possibilities although it is unlikely that any one central reprographic unit will want to undertake all of them. They do, however, give some indication of the variety of work from which particular tasks can be chosen.

Stencil duplication is, even today, more restricted than small-offset. On some modern machines the variety of papers which can be used is wider than formerly but they still have to be reasonably absorbent. The electronic stencil cutter and more recently the heat process stencil allow virtually any original to be copied, and there is even one electronic stencil cutter which enables process colour to be copied. However, I have not seen any colour copies made in this way which come up to small-offset standards. The masters need to be made by a stencil cutting service and this can be quite expensive, although considerably less than a set of small-offset process colour plates.

However, when used for straightforward duplication, the running costs of a stencil duplicator are higher than when the equivalent work is turned out by small-offset and, provided the small-offset is carefully used, the quality of stencil duplication is not so high. To compensate for this the cost of equipment is lower as is

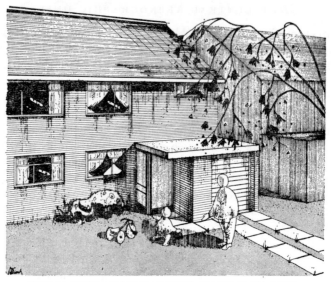

BUILT WITH THE MODERN FAMILY IN MIND

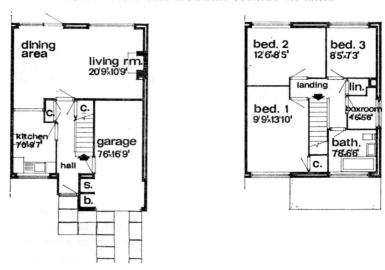

GROUND FLOOR

FIRST FLOOR

These dimensions are approximate

Fig. 9. An example of work which can be printed on a small-offset machine. This, and **Fig. 13**, were originally in two colours. Note that line drawings have been used for illustration in both cases. This eliminates the need for screening and the plates could therefore be prepared on equipment available to the average internal print department.

the degree of experience needed to operate it well. Hence, other factors being equal, it may be possible to save the extra materials cost on wages.

Fig. 10. A further example of work printed on a small-offset machine.

The place of spirit duplication in a central copying unit is arguable. Detractors of the process refer to it as 'that purple mess' but, as I have already pointed out, this is not necessarily so provided reasonable precautions are taken. If there is a consistent demand for from say 25 to 250 copies of a readable standard which must be produced as cheaply as possible it certainly has claims to consideration. With care, and using black hectographic carbons well within their exhaustion limits, a quality approaching stencil is possible but the process is best considered on grounds of cost and simplicity. Consider it for any jobs where cost rather than appearance is the primary consideration.

Even now it is a little difficult to be quite objective about xerography as a duplicating process. The equipment is available through only one company and their method of charging differs from that imposed for any alternative equipment. The method of charging on a metered basis for each copy made means that you know how much the basic cost per copy will be. Do not overlook the fact, however, that this does not include the cost of paper or developer powder. If you use this kind of equipment as a duplicator so that the majority of copies are made at the lowest rate, some users claim that adding 25 per cent is more than sufficient to cover the cost of consumables plus the monthly rent. Others say that these factors can be as high as 50 per cent of the metered cost.

Quite a lot seems to depend on how the machine is used. However, the data is available from the manufacturers and potential users who take the trouble to work out the best and worst case will find that their actual costs fall somewhere in between.

Undoubtedly the fact that no special master is needed and that the original can be prepared on plain paper contributes to overall speed of operation as does the fact that 'makeready' is reduced to replacing the copied master with a fresh one. Because of this, it is probable that where the average run is short xerography is faster than other duplicating methods, as has been claimed for it. Overlays can also contribute to speed of operation as they mean that any standard information used on a number of masters need only be prepared once. In addition they enable titling and logos to be incorporated in the copy. This means in effect that the only practical restrictions are that everything reproduces in black and that illustrations should be confined to line work. Further, a xerographic copier/duplicator will only feed one weight of paper although they can be adjusted to take different weights up to and including card. Small originals can be copied two or four up and cut afterwards. Assuming that these restrictions are acceptable there is no reason why xerographic copier/duplicators should not be used in a central reprographic unit or that the overall cost need be substantially higher than when using some other processes.

As I pointed out earlier there is no one ideal method of copying or duplicating which will meet all situations, nor will the range of work undertaken internally be the same for each company. The options available can be summarized as follows:

1 For straightforward duplicating where the average run is, say, from 25 to 250:
 a Spirit duplication is the cheapest to install and to operate. The quality of copies is not so good as with other processes and considerable self-discipline has to be exercised if 'the purple mess' is to be avoided.
 b Xerographic copier/duplicators. The cost per copy is considerably higher than with spirit but operation is extremely clean and simple and there are indications from users that the extra operating costs can be made up by increased production.

If nothing more is wanted it is a straight choice between these two processes but by definition a central reprographic unit which

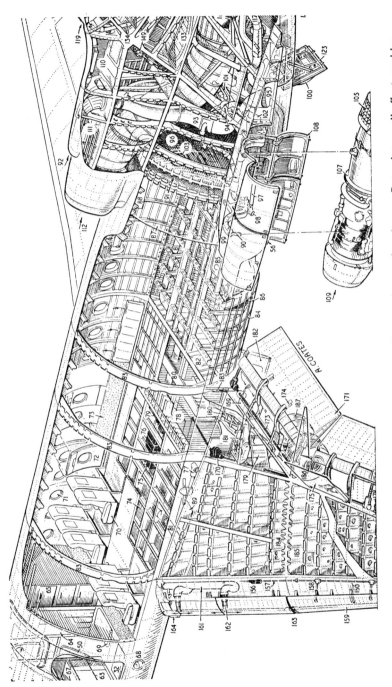

Fig. 11. This complex cutaway diagram of the Trident 1E was also produced on a Rotaprint small-offset machine.

undertook only this work internally would, in effect, be a print buying department.

2 For longer run work the following should be considered:

a Small-offset which has the flexibility and versatility to tackle virtually any print job required. However, all kinds of small-offset work cannot be undertaken economically on the same machine. To a large extent the variety of work which you can undertake economically yourself is governed by the volume of each kind and consequently the amount of experience which you can command. The subject is too complex to review here in any detail but generally speaking you have the following options:

(i) You can keep to small format machines (i.e. just over foolscap) and if the volume of work increases sufficiently add more machines.

(ii) You can have a large format machine (up to about $17\frac{1}{2}$ by $22\frac{1}{2}$ in sheet size) or any mixture of small and large format machines which suits your workload.

(iii) You can undertake only straightforward work on 'direct image' plates internally from copy preparation to completion. You can then go to an outside platemaker for all plates which cannot be 'made' on 'direct image' plates.

(iv) You can keep to small format but equip yourself to undertake platemaking internally by one or more of the methods previously described. This is usually most successful where the primary requirement is for overall speed. If you want to undertake platemaking by traditional methods in competition with a trade platemaker you need a large volume of work to justify the extra semi-skilled staff, space and cost of equipment.

(v) You can also install various kinds of finishing accessories which are discussed later.

b Stencil duplication may be justified on the grounds of lower capital investment and the lower level of experience needed if there are adequate outside print services which are willing to undertake your short run work (under 1,000 copies) quickly, reliably and on time. As previously mentioned, the range of originals which can be duplicated has been extended in recent years by the introduction of electronic stencil cutters and heat

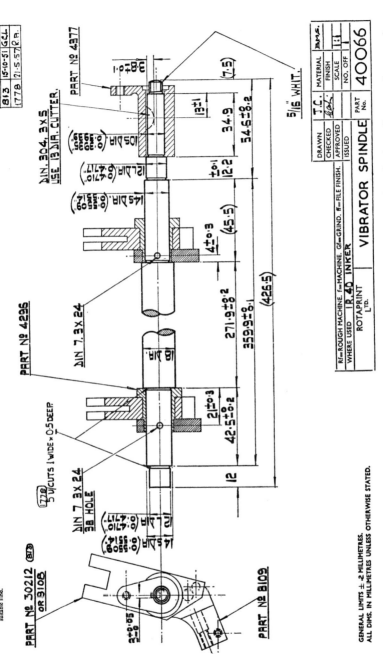

	MOD N?	DATE	SIG
	659	8.2.51	T.C.
	813	15-10-51	G.C.
	1778	21.5.57	P.B.

PART N? 4377

PART N? 4296

DIN 304. 3×5
USE 13 DIA. CUTTER.

DIN 7. 3×24

DIN 7. 3×24
3B HOLE

5 V/CUTS 1 WIDE × 0.5 DEEP.

1778

PART N? 30212   813
OR 9108

PART N? 8109

13±1

34.9

54.5±0.2

12.2 ±0.1

(45.5)

4±0.3

271.9±0.2

359.9±0.1

(426.5)

2±0.3

42.5±0

12

3±0.05

18±0.1

(7.5)

5/16" WHIT.

GENERAL LIMITS ±.2 MILLIMETRES.
ALL DIMS. IN MILLIMETRES UNLESS OTHERWISE STATED.

DRAWN	T.C.		MATERIAL	B.M.S.	
CHECKED			FINISH		
APPROVED			SCALE	1:1	
ISSUED			NO. OFF		

Rf=ROUGH MACHINE. f=MACHINE. Gf=GRIND. ff=FILE FINISH.

WHERE USED   I R.40   INKER

ROTAPRINT LTD.   VIBRATOR SPINDLE   PART No.   40066

Fig. 12. An engineering drawing printed small-offset on to translucent material so that further diazo copies can be taken. This technique is useful where comparatively large numbers of drawings have to be supplied either to subcontractors or to associated manufacturing facilities where microfilm facilities are not available.

process stencils. The other limitations already mentioned remain.

Largely due to the spread of small-offset a variety of accessories have become available over the last decade which simplify the finishing of duplicated work, enabling it to be undertaken in a central reprographic unit:

1 A wide range of automatic and semi-automatic collators is available from simple hand-operated machines with less than 10 stations up to high speed automatic models capable of sorting 50 or more originals at very high speeds. Even the small hand operated machines are several times faster than collating by hand, a job which can take a good deal of labour from other tasks at peak periods. The more sophisticated models, on the other hand, although costing hundreds of pounds can save their initial cost where there is a sufficient volume of suitable work. A comparison of the relative specifications will indicate the volume of work of which various models are capable. Automatic collators can be divided into two groups, flatbed, where all the stations are on the same plane, and the rotary drum type. Some of the latter type are designed to be attached to the copy tray of a duplicator, to enable work to be printed and collated in one operation. A third type which consists of tiers of pidgeon holes with a device such as a friction wheel to push the top sheet forward is usually only semi-automatic, although completely automatic machines have been designed round this system.

2 Collated sets of sheets can be punched and made up into looseleaf booklets by using one of the several types of binding apparatus available. These consist of a punch and 'comb' which opens a plastic spine and inserts the teeth of the spine through the punched holes. This kind of binding is particularly suitable for service manuals and sales kits where a degree of permanency has to be combined with the ability to change pages with comparative ease. A much simpler type of binding consists of a plastic tube cut along its length and slightly opened along the cut. When inserted over the spine of the sheets friction keeps them in place. No apparatus is needed but the papers will not, of course, open flat.

3 As an alternative, adhesive binding can be undertaken by using one of the simplified systems now available. The secret of these

is in the materials, the apparatus consisting only of a 'jogger' to get the paper into exact alignment and a clamp to keep them there while the adhesive is applied to the spine. An infrared lamp is sometimes provided to aid the drying-out process. If required the 'book' can then be finished off by applying self-adhesive binding tape. Similar adhesive materials can be obtained for padding. These adhesives, usually of the latex type, are weaker so that sheets can be torn off the set. A padding press can be used but is not strictly necessary. Padding, although primarily intended for making up sets of printed forms, can be used to use up offcuts turning them into scrap or memo pads.

4 Even where most of the paper is bought already cut to size a guillotine is a useful accessory. With it forms and other papers can be trimmed and scrap of which only a small part is printed can be cut to a smaller size and reused. Where work is printed two or more to face or paper bought in mill sizes it is essential. A variety of models is available, from power-operated machines capable of cutting mill sheets 500 at a time down to hand-operated trimmers which handle perhaps 24 or so sheets of average office stationery weight.

Preparing the copy

The preparation of the copy may take place either in the central reprographic unit or in the department where it originates. There is no hard and fast rule. On balance it is probably better to have an art studio attached to the reprographic unit than elsewhere as not only can its services be shared by several departments but those working in it are more likely to appreciate the possibilities and limitations of the reprographic equipment in use. However, there may be departments such as those dealing with publicity or technical illustrations where it is more economical to have a separate studio attached to the department. These can then be considered as a 'remote input terminal' for the central reprographic unit and close liaison be established to ensure that the reprographic requirements of the artwork produced are met. This is important if time is not to be wasted on artwork which cannot be reproduced with the equipment available when, perhaps, a small alteration in the concept would have made it acceptable, or less expensive to reproduce.

Any commercial studio preparing work for reproduction by

duplicating rather than printing methods must bear these limitations in mind, even when small-offset is to be used.

Unless small-offset plates are to be made by a trade platemaker or a fully equipped platemaking unit is available internally, it is better to avoid photographic or tone illustrations. These require 'screening' which can be accomplished by some copying methods usable by an internal print department, but if used, much of the basic simplicity is lost.

On the other hand, virtually all line work can be transferred to a small-offset plate or other duplicating master. Where the office copying methods are in use it is preferable for the line work to be bold and simple, particularly if the person making the plates does not have the kind of expertise which comes with doing the job as a full-time occupation.

There are a number of ways in which copy can be prepared, some of which require only the degree of skill which can be acquired by a person without specialist training in commercial art.

For titling and other display work adhesive transfer lettering can be used, but not on direct image plates unless it has been specially made for this purpose. At the time of writing only one adhesive transfer lettering had been marketed for this. Most of the cold type composing machines based on the normal typewriter can be used by a trained typist after a short period of further instruction. The length of this period varies as does the ability of these machines to vary the typeface and/or type size. Of the two systems which allow some variation, the IBM is based on the Model 72 'golf ball' typewriter, the golf-ball typing-heads being available with a number of different typefaces, including technical and mathematical symbols and foreign languages. On the simpler machine justification of the right-hand margin entails a second typing but a more expensive version enables this to be done automatically from a magnetic tape recording of the original typing. Olivetti market a typewriter for making direct image plates on which the number of 'strokes' can be stored by pressing the keys while the typing bars are inoperative. It is not, however, possible to change the typeface on this machine.

The Varityper range uses type segments which can be changed at will. Different models in the range have varying capabilities, some being made for right-hand margin justification and others for primarily columnar work.

Where type variation is not required a proportional spacing typewriter can be used. If necessary, right-hand justification can be achieved by a second typing after the line width on the first typing has been kept within predetermined limits.

Mechanical tints are more difficult to use than transfer lettering but are not beyond the skill of anyone with an aptitude for this kind of work who is prepared to take considerable care. In addition, there are today a number of preprinted figures and stylized drawings made on the same principle as adhesive transfer lettering. If anyone without commercial art training is to try illustrating, these are easier to use than other methods. Do, however, keep the layout simple and straightforward. Elaborate layouts require skill and experience if they are not to look amateurish.

By using one or other of the indirect methods of making masters the methods described above for making originals can be used for any of the main duplication processes. One piece of advice applies to all layouts—keep them bold and simple unless you have the skill, art training and experience to do more advanced work, and the equipment to reproduce it successfully.

Management considerations

Do not overlook the fact that a central reprographic unit is the hub of a visible communications network. Even where the more urgent systems work is undertaken in the offices which the system serves, as it should be, the central reprographic unit should be located where it has good and frequent communications with all departments in your organization. When planning a messenger service or other internal communications the needs of the reprographic unit should be studied. In some instances where there is a heavy demand for small numbers of copies and the copying service is highly centralized, it may even be worthwhile to install an air tube system to convey originals to the central reprographic unit and to return the copies. However, such systems are quite expensive to install and it is doubtful if even this 'form of communication' would make systems work as efficient when undertaken in a central unit as when done in the office to which the system applies.

However the work is distributed, it should be costed according to the normal tenets of cost accounting and a strict watch kept on the cost of the reprographic service as a whole. The costing of the complete service, regardless of where the machines are located, should be the responsibility of one person, most conveniently the communications manager. If this is not done the cost of the service will get out of hand as it is difficult to amalgamate costings which are derived from a number of sources.

It is, of course, the cost of the overall service which is important and once this has been ascertained, not only for the central reprographic unit but for the 'peripherals' as well, an estimate of the 'cost effect' on related procedures should be made to ensure that the copying service is fulfilling its function as economically as possible. The 'cost effect' is not easy to gauge and frequently involves estimating projected losses which would have occurred if a less efficient system had been employed. For example, if by using a reprographic method the monthly statements are posted on the first of each month rather than on the fifteenth, it is reasonable to assume that a saving will be made equivalent to the extra interest which does not have to be paid on the outstanding money over this period. The saving will not be exact as it is impossible to estimate the exact amount which has been paid promptly because the statements were sent out promptly. However, an approximation based on past experience is possible.

CHAPTER 18

The legal aspects

The legal aspects concerned with photocopying or duplicating can be divided into two categories, copyright and acceptability. This latter applies to certain documents which are legally binding, some of which affect only lawyers. Of the two copyright is frequently misinterpreted and I will deal with this first.

Copyright and copying
The relevant Act which deals with the position of those who wish to copy part or the whole of a copyright work is the Copyright Act 1956 and any readers who have doubts as to the legality of any copying they wish to undertake should obtain a copy of this Act and consult a lawyer who is conversant with its provisions. I know it is easy to say 'consult a lawyer' and that this can result in paying out considerable sums in legal fees. In cases of doubt, however, it is cheaper to consult a lawyer before the event than to have to call one in to defend your action afterwards. This is particularly true of a copyright originating in some foreign countries where the legal climate may be more pedantic than in Britain and less anxious to be be fair to both parties.

This much said, so far as the United Kingdom is concerned, the provisions of the Copyright Act, so far as they apply to copying of copyright material are as follows :

1 No infringement of copyright can arise unless a substantial part of the copyright work is involved.
2 Even where a substantial part is involved, such copying may be permissible as fair dealing provided :
a It is for the purpose of research or private study (Section 6 (1)) :
b It is for the purpose of published criticism or review (Section 6(2)).

As is fairly common in British law no attempt is made to define 'substantial part' since, as to try to reach a definition of this term which would be relevant in all circumstances is well nigh impossible and would only lead to inconsistencies and probably injustice. However, this term is of prime importance since it may be taken in law as overriding any other limitations contained in the Act. It is reasonable to suppose that a 'substantial part' of any copyright work must depend on :

1 The length of the work as a whole.
2 The importance of the extracts in relation to the work as a whole.

For example, a claim that the extract was not 'substantial' could hardly succeed if the extract included a summary of the work, or of a part of the work dealing with a substantial subject, either in words or, for instance, in the form of a diagram, graph, statistical or mathematical table or scientific formula, so that the core of the work, or any specific part of it, was contained in the extract.

To be more specific : if, for instance, you have a marketing report covering a range of domestic electrical products and you are interested in marketing washing machines, assuming that the report is the copyright of the compiler, it would probably be an infringement of copyright if you extracted all the information relevant to the marketing of washing machines even if you did not copy information relevant to any other product. On the other hand if you copied a part of the information only as it referred to a number of products, for instance, area distribution of one product, price ranges of another and technical details of a third, this would probably not be an infringement of copyright even if the extracts were together longer than all the information contained in the report relevant to washing machines. If you extracted all the tables, graphs, etc, from the report this would be an infringement provided they were sufficient to give you the substance of the report. It should be borne in mind that, generally, if something is worth copying it has substance and is thus likely to infringe the copyright.

It would appear that what the writers of the Copyright Act had in mind in using the term 'substantial part' was a part which would be useful to the person making the copy, that is, a part which he could use in place of the original document, thus depriving the copyright holder from benefiting from his copyright by depriving him of the sale of one or more copies of the original work which the person making the copy would otherwise have bought. This, after all, is fair and reasonable. Suppose, for instance, that you are interested only in microfilm. If you borrow a copy of this book and copy the chapter on microfilm, after which you return the book to whoever you have borrowed it from, you are depriving the publisher of the profit he would have made on a copy and me of my royalty. This, therefore, is a substantial part of the book. On the other hand, if you copied the first page of each chapter with a view to showing this extract to your executives to find out if the book would be useful to them, this would not be a substantial part of the whole, although the extract would be longer than the chapter on microfilm. In the first case the person making the copy would benefit at the expense of the publisher and author whereas in the second case he would not.

We now come to the question of making extracts for the purpose of research or private study. Section 6(1) of the Act states that 'no fair dealing with a literary, dramatic or musical work for the purpose of research or private study shall constitute an infringement of the copyright of the work'. Legal precedent, however, has established that such 'fair dealing' must be exercised by the individual for himself, and not by any one person on behalf of another. Consequently, a student or research worker may make copies of extracts from a work, or even copy a complete work, for himself; he may not make such a copy for another without, in effect, publishing the work and thus infringing the copyright.

It would, therefore, appear that the copy must be made by the person who is to use it, but it is doubtful if this is intended to mean that the user must actually operate the copying machine himself.

What it does mean, however, is that one research worker cannot legally copy a published document for a colleague. If he does this he may have infringed the copyright. Before passing on to the position of librarians it should be noted that this applies only to copies made for research or private study and applies only to copies made for the personal use of the person making the copy. It does not apply to copies made in the furtherance of industry or commerce as such. For instance, if an executive makes a number of copies of a published piece of research so that several colleagues can study it simultaneously and distributes these to his colleagues this would not strictly speaking be fair dealing within the meaning of the Act, unless permission could be implied from surrounding circumstances. For example, if the company had already approached the proprietor of the copyright with a view to backing further research along the same lines this would probably be viewed quite differently from a situation where the company intended gaining advantage from the published research without further reference to the proprietor of the copyright. However, we are now getting into a situation where something more than copyright may be involved such as patents law which I do not intend to discuss in this context. The point to remember here is that if you want to copy the whole or a substantial part of a work this can only be done legally if the copy is for your own use. Although not of so much direct interest in business it is worth mentioning the position regarding criticism and review before going on to the position of librarians.

I have already mentioned that the copying of a substantial part of a copyright work may be permissible for the purpose of criticism or review. In view of the fact that a 'substantial part' is not defined in the Act, the Society of Authors and the Publishers Association jointly agreed that it would be convenient if there were to be some guideline which would apply in general practice. Consequently they agreed on the following formula :

'. . . would not regard it as "unfair" if for purposes of criticism or review a single extract up to 400 words, or a series of extracts (of which none exceeded 300 words) to a total of 800 words were to be taken from prose copyright works, provided that in no case this amounted to more than one quarter.'

This joint statement refers specifically to the provisions of Section 6(2) which relates to 'fair dealing' for the purposes of criticism or review. The Act requires that any such use shall be accompanied by sufficient acknowledgement.

Although such uses affect primarily reviewers in commercially published books, magazines and newspapers, the Act also applies to such publications as company magazines, newsletters and other prestige/promotional material. In this context it could apply to work done by a duplicating department or internal print shop.

Photocopying and librarians
The rights of libraries to make photocopies or duplicates of copyright works depend to some extent on the purpose for which the

library exists. The special exceptions laid out in Section 7 of the Copyright Act 1956 apply only to those libraries which are prescribed by the DTI under regulations made under this Section of the Act, known as the Copyright (libraries) Regulations 1957, which define the libraries entitled to take advantage of these exceptional provisions as those operating in schools or establishments of further education covered by Section 41 of the Education Act 1944 (and by Section 39 of the Northern Ireland Education Act 1947 and Section 143 of the Scottish Education Act 1946), as well as any public library or government department library or any library which exists for the purpose of facilitating or encouraging the study of all or any of the following: religion, philosophy, science, (including any natural or social science), technology, medicine, history, literature, languages, education, bibliography, fine arts, music or law. Of these, the subjects most likely to be found in a library concerned with industry or commerce are science, technology, medicine, languages and law.

Any library which is established or conducted for profit is explicitly excluded from these privileges, even if it covers only one or more of the subjects mentioned. For instance, a library run by an association of technically or professionally qualified people for the benefit of its members would qualify as a privileged library provided it was not run for profit. A similar library run as a profit making concern would not.

Even so, privileged libraries are not entitled to make any number of copies for any or every person who asks for one.

The libraries covered by the Department of Trade and Industry Regulations by virtue of Section 7 of the Act are free to make single (not multiple) copies of:

1 Articles from periodical publications provided no person is supplied with a copy of more than one article from any one publication. It should be noted that more than one copy may be made at one time provided the above restrictions as to distribution are made. You can, for example, make several copies of the same article or copies of several articles from the same periodical. Further, it is generally assumed that the term 'periodicals' includes annual reviews and year books for this purpose.

2 Extracts from copyright books, either where the copyright

owner has given permission or where it is impossible to trace him and the librarian is satisfied that every endeavour has been made to do so. In this connection it should be noted that an undertaking given by the person asking for the copy to the effect that either permission has been obtained to make the copy or that the copyright owner cannot be traced does not protect the librarian in law. The onus of obtaining permission is on the librarian to obtain such permission where, to quote the Act, 'the librarian knows the name and address of a person entitled to authorize the making of the copy or could by reasonable enquiry ascertain the name and address of such a person.'

3 In the case of both periodicals and extracts from books the Regulations lay down that single copies may be made by the privileged libraries only for the purpose of research and private study for those who have 'not previously been supplied with a copy . . . by any librarian.' The libraries are required to furnish from the person requesting such copies a formal declaration and undertaking in writing to this effect.

4 The privileged libraries allowed to supply such copies are required by the Regulations to charge 'not less than the cost, including a contribution to the general expenses of the library, attributable to their production.'

It follows, therefore, that any library which allows the free and unrestricted use of copying equipment on its premises is liable to put itself in a position where it may be contravening the Act and Regulations. To the best of my knowledge the exact responsibility regarding copies made on a copier which is used by the public has not been tested in the courts. There can be little doubt that the user must still take responsibility for the copies he makes whether on his own or someone else's machine.

The position of a library run by a company for its own benefit appears to be a little obscure. To the best of my knowledge the status of a library which exists solely to meet the needs of a company has not been challenged in the courts. On the one hand such a library is usually in existence to 'facilitate or encourage the study of science, (including natural or social science), technology, medicine or law' (as it applies to a particular kind of industry or business). On the other hand, although the library itself is an overhead its purpose is undoubtedly to contribute indirectly to the profits of the parent company.

In practice a company library frequently does copy either extracts or even complete copyright works and provided such copies are for the sole use of members of the staff there would appear to be little possibility of copyright proprietors being in a position to test the validity of this procedure. Where copies are taken of articles from controlled circulation magazines no harm would appear to be done. The publisher is, in effect, giving away his copyright by distributing the magazine free of charge. However, where a 'reader' makes copies of the whole or part of a periodical, book or other work which is charged for, he is in fact avoiding paying for extra copies. It is, therefore, the obvious intention of the Copyright Act that he should not do so, even if these are for the use of his own employees. It would appear to be the intention of the Act that by making copies of this kind the person who does so is committing a breach of copyright. The fact that it is a breach which is difficult to detect does not make it any less of a breach, nor does it put the person who makes the copies on the right side of the law.

However, there are circumstances in which copying may be permissible as 'fair dealing . . . for the purpose of research or private study' under Section 6(1). To cover such situations the Society of Authors and the Publishers Association have suggested limits which authors and publishers generally would not regard as unreasonable for photocopying purposes. They have agreed:

> That they would not normally regard it as unfair if a single copy is made from a copyright work of a single extract not 4,000 words, or a series of extracts (of which none exceeds 3,000 words) to a total of 8,000 words, provided that in no case the total amount copied exceeds 10% of the whole work. In the case of short works each complete work must be considered on its own when judging the maximum of 10% of the whole and not the volume into which it may be bound with other works.

It should be noted that this is not a matter of law but only an indication by two bodies representing the generality of copyright owners of what they would consider reasonable. It is, however, probable that if a court of law had to decide what is reasonable this joint statement would be taken into consideration. The important issue from the point of view of company libraries is that this general permission is intended to apply to all libraries and not just to the ones covered by Section 7 of the Act and the Regulations attached to it, i.e. government and public libraries plus those existing to further one or more of the arts or sciences enumerated.

This in fact sets out a guideline as to what is reasonable for purposes of research or private study under Section 6(1) of the Act where no permission is required to make a single copy. It does not relieve the librarian from obtaining the copyright holder's permission where 'substantial parts' are copied under Section 7.

So far as the copying of illustrations is concerned the Society of Authors and the Publishers Association have included the making of single photo-positives or slides from illustrations for the purpose of providing an illustration which can be projected for purposes of instruction in a school or other educational establishment. Now that education in the broader sense of the word has come right into industry the point as to whether classes held by a company as part of an industrial training programmme would be regarded as part of an 'other educational establishment' is one which so far as I am aware has not yet been decided. In any case to make multiple copies of such illustrations would be a clear breach of copyright and permission must always be sought before this is done. This in fact applies to the making of multiple copies from any copyright works whether in whole or in part, even if the part is not substantial within the meaning of the Act.

It is not, quite obviously, the intention of the Act that scholars should individually ask for copies of a copyright work so that collectively each member of a class has a copy which can be used for class studies. Class study would not be considered to be private study. Even less is it the intention of the Act that companies should 'get round' the Act by individual members obtaining copies of copyright works for the purposes of collective discussion or study.

What is a photocopy?

Quite briefly a photocopy is a copy produced by any photographic, electrophotographic or mechanical means. The Act is intended to protect the interests of copyright owners so that publishing will

continue to be economically viable. The means by which the copying is done is not important and it is quite obviously not the in-
tentions of the legislators that any system of copying should be
used as an excuse for not complying with the provisions of the
protection afforded to copyright owners simply because on technical grounds the process could be argued as not being based on
photography. Nor would microfilm be excepted simply because the
initial image is too small to be read with the naked eye. In short,
the process used to make the copy is immaterial, a 'duplicating or
photocopying process' including virtually any kind of process involving the use of apparatus for reproducing multiple copies.

What can you copy?
The international copyright mark, usually accompanied by text
defining the copyright owner, will be found in every publication
for which copyright is claimed, except that in older publications
the international copyright mark may be replaced by the word
'copyright'. This mark is normally found on the verso title page
in a bound volume and on the contents page in a periodical. Where
copyright is claimed a businessman or employee acting on behalf
of his employer can:

> Make single copies of a part of a copyright work, provided
> that this is not a substantial part, that is to say provided the
> core of the work (or any specific part of the work) is not
> contained in the extract.

As previously discussed the Society of Authors and the Publishers
Association have agreed that they would not normally regard it as
'unfair' if a single copy is made of an extract 'not exceeding 4,000
words or a series of extracts, of which none exceeds 3,000 words,
to a total of 8,000 words, provided that in no case the total exceeds
10% of the whole work'.

It is the obvious intention of the Act that any copies made should
not be such as to avoid the necessity to buy extra copies of the
work or where the copy from which the extract has been made
is borrowed to buy a single copy. In other words if you copy from
a copyright work with the intention of depriving the publisher of
the sale of one or more copies you are acting against the spirit of
the Copyright Act and do so at your own risk.

The exceptions to this rule outlined above are intended to

facilitate research and private study and can be claimed only by certain categories of librarians as defined earlier in this chapter.

How acceptable are copies in law?

To be acceptable in a law a copy must be permanent, at least for the vast majority of uses. Since there are a large number of such uses, many of which are seldom met with in the general run of business, I will deal with only those most commonly used.

I PATENTS

The United Kingdom Patents Rules (Rule 5) states that all documents and copies of documents, except drawings filed at the Patent Office should be written, typewritten, lithographed or printed in the English language in legible characters with a dark indelible ink. It is separately stated that copy documents may be carbon copies provided that they are on paper of good quality and that the typing is black and distinct.

From the above it is clear that copies made on any offset lithographic machine are acceptable, including small-offset, since there is no basic difference in the process, and when these are well made it is difficult to tell what type of offset lithographic machine has been used.

The Patent Office has not so far made any statement on the acceptability of electrophotographic or photographic copies. Xerographic copies are usually accepted provided that they have been properly fused and it would appear that copies made by other processes are also accepted in practice provided that they are permanent and legible. Acceptance of such copies would, however, appear to be at the discretion of the Patent Office.

2 DOCUMENTS WHICH MUST BE PRINTED

The Companies Act 1948 requires the following documents to be printed :

1 Articles of Association.
2 Altered Memorandum of Association.
3 Ordinary Resolutions increasing the capital of any company.
4 Special and Extraordinary Resolutions of Public Companies and non-exempt private companies, and copy agreements filed by such Companies under Section 43 of the Act.

The Registrar will accept that this stipulation has been satisfied by the following processes:

Letterpress, gravure, lithography
'Office typeset'
Offset lithography
Heat fused electrostatic printing, but not other photographic processes or office copying methods
Stencil duplicating, but not spirit duplicating

All this is provided that in general appearance, legibility, format and durability the document is suitable for publication and use on the files of Companies produced by the Registrar for public inspection.

Where the latter two categories of documents are concerned (i.e. 3 and 4) a 'copy produced in some other form approved by the Registrar' will be acceptable under the Companies Act 1967. In practice this leaves it open to the Registrar to accept copies made by any copying process which fulfils his requirements as to general appearance, legibility, format and durability.

It should be noted, however, that except where documents are letterpress printed they must be endorsed with or accompanied by a certificate by the printer stating the process used.

It is difficult to find evidence as to how permanent copies are produced which have been made by various processes. This depends to some extent on how well the copy was made in the first place and with some processes on the conditions of storage. However, some considerations are:

1 Microfilm is as permanent as any other photographic film. It is a process based on silver halide materials and under reasonable conditions of storage can be considered as being permanent.
2 Xerographic copies can also be considered as being as permanent as the paper on which they are printed. This does not appear to have been tested in this country but they are accepted as permanent records by the US Patent Office and their permanence is attested by the United States Testing Company Inc. (Report dated 11 May 1960) and by the National Institute for Materials Testing, Stockholm, Sweden (Certificate No. U64-3754 dated 15 October 1964).

3 Stencil duplicating can be considered to be permanent. The paper used is absorbent and the ink in many ways similar to to a printing ink. Hence the image should last as long as the paper.

4 Small- or office-offset is as durable as any other form of lithography, the image when properly applied being as permanent as the paper.

5 The permanence of electrostatic copies other than xerographic depends on how well they are fused. This varies from one model to another. Where permanence is important prospective buyers should obtain proof of permanence before buying as in the present state of development it is impossible to generalize.

6 Copies made by the chemical photocopying processes are roughly as permanent as photographic prints, permanency depending on how thoroughly they have been made and on the conditions under which they are stored.

Some landmarks in the history of microfilming

The earliest known micro-images were made by an English photographer called Dancer in about 1835 by attaching a camera to a microscope. In this way he achieved a linear reduction of 1 : 160, a slightly greater reduction, incidentally, than PCMI. However, this seems to have been regarded as a technical curiosity at the time and the first commercial application, at considerably lower reduction, did not take place until 1875 when the archives of Rothchilds Bank in Paris were microfilmed.

In the meantime microfilm had been used during the siege of Paris, 1870–71, when microcopies measuring only 4 by 6 mm were made of documents up to 500 by 700 mm. These microcopies were presumably smuggled out of the besieged city. At the same time another photographer was making collodian microcopies at a reduction which enabled several thousand words to be carried by the famous pigeon post which operated during the siege.

Other uses up to this time for micro-images had been merely in novelties, dominoes much like those offered by souvenir shops today, which enabled local views to be seen by holding the domino up to the eye. However, it was not until 1910 that two Belgians, Messrs Goldschmidt and Paul Otlet, introduced their 'Bibliophoto' at a conference. This was probably the first use of transparent microfiche and opened the way for documentary microfilm as we know it today. Military applications received a fillip during the First World War, 1914–18, microfilm being used extensively by both sides, mostly for spying.

It was not until 1926 that the 'Checkograph' was designed. This can be said to be the start of business applications in the microfilm field. In fact it was not until nearly two years later, in 1928, that a New York bank installed a system with the idea of preventing cheque frauds which were prevalent at that period.

After that the application of microfilm in commerce and industry grew fairly slowly until the outbreak of the Second World War. This time espionage accounted for only a small part of the applications. Numbers of London business houses had important documents copied, including the banks, as a protection against losing essential information by enemy action during the blitz. The American Army used microfilm to handle the home mail of soldiers in the field and even in the USA several civilian organizations started to use microforms, although there space saving and ease of transport were more important than the rather remote possibility of originals being destroyed by enemy action.

The postwar period did not see the anticipated growth in the use of microfilm which had been so confidently predicted during hostilities. This was partly due to shortage of materials and equipment, and partly due to the fact that most people still saw microfilm as a means of storing information in a safe place.

It was not until the information explosion of the fifties and sixties had increased many files to unmanageable proportions that people have started in recent years to turn once again to microfilm as a solution, this time as active systems.

Glossary of microfilm terms

Note: This glossary does not include normal photographic terms which are used in the same sense in microphotography as in general photography.

ACTIVE : (MICROFILM SYSTEMS)   In business this term is applied to any system where the information recorded is intended to be acted upon. Either the administrative routine is incomplete at the time when the information is recorded, or the information affects the present running of the business. (See also Archival.)

In library work a microfilm system is said to be active when the microfilm copy is intended to be used in place of the original work. For example, where a sales documentation routine is microfilmed before the accounts department has received the money for the transactions recorded, the system is active in the sense that the overall office routine is incomplete. The recording has been made in the anticipation that it will be used either in the event of an invoice being queried or in the event of a customer sending in a repeat order without detailing the goods required. A management report microfilmed at the time when the information becomes available is active in the sense that the purpose is to enable executives to study the microfilm copy and act on it.

In a technical library the report of a current development may be microfilmed so that the company's engineers can study and act on the report or the institution doing the original research may have distributed their finding in the form of microfilm.

ARCHIVAL (MICROFILM)   Microfilm is said to be archival when the information recorded on it has only historical value or, in business, where it is kept only in case the information should be needed, the implication being that in normal circumstances it will not be. For example, in the early days of the Second World War a number of companies microfilmed their essential records and stored them in a safe place in case the originals were destroyed by enemy action.

APERTURE, CARD   A card in which an aperture is cut so that a single frame of microfilm can be mounted in it. The standard aperture card is an 80-column punch card in which a frame of 35 mm film is mounted. Forty columns are then still available to punch a code enabling the information filed in this way to be sorted.

The system is usually used to record engineering drawings, but could be used to record oversized business documents. In the past there have been a number of unorthodox aperture cards available but while it is tempting to use 'odd' cards for special applications it is preferable to use the standard system as supplies are more readily available. The facility for sorting aperture cards on a punched card sorter is not so frequently used as was anticipated at one time. When used dummy cards are normally sorted in this way, and the sorted cards then pulled by hand from the file.

CAMERA   (See Flow Camera and Planetary Camera.)

CASSETTE   A special kind of enclosed reel in which roll microfilm is mounted. The purpose of the cassette is to enable the film to be taken up automatically by the threading device on a reader or reader/printer equipped to take cassetted film.

COM (COMPUTER OUTPUT MICROFILM)   Any system whereby the output from a computer can be recorded directly on to microfilm without an intermediate printout on to a conventional computer output device, such as a line printer, is described as COM. (See text on page 100.)

DIAZO (FILM)   A special type of film used for making copies of an original microfilm. Diazo film is of the heat development type and hence needs different processing to a normal silver halide film. Development takes place in seconds, using an automatic processor or duplicator unit designed for diazo film.

DRY SILVER   A proprietary system marketed by the 3M company. The system, which uses a form of heat development, enables the film to be developed or a hard copy printout to be made in seconds rather than in minutes and without removing the film or material from the camera or reader/printer respectively. However, the equipment has to be made for the system.

FICHE   Literally French for card. In microfilm a fiche is a rectangle of film on which a number of frames are recorded in rows running across the longer dimension of the rectangle. A space is left at the top of the fiche in which the title is recorded in letters large enough to be read by the naked eye.

The two standard fiche sizes are 4 by 6 in and 3 by 5 in. In the standard layout recommended by BS 4187 the larger of these sizes (actually 105 by 148 mm) are five rows each of 12 frames and in the smaller size three rows of 10 frames, making 60 and 30 images per fiche respectively. Each single frame is 11·25 by 16 mm which is the same as a standard frame on 16 mm roll film.

Fiches can be made either by filming on to an original fiche using a special camera or by filming on to 16 mm roll film and cutting the film into strips of the appropriate length which are then stuck to a clear acetate base. The former gives the best possible quality fiches. Recently there has been a tendency to reduce the frame size further than the recommended standard, enabling a larger number of images to be recorded on a standard fiche, 98 in some systems and over 200 in a few.

FLOW CAMERA   A camera designed for the high-speed recording of documents on to microfilm. The documents are passed across a roll of film as it is advanced from one frame to the next, the speeds of document transport and film being synchronized. Apart from speed of operation, up to 100 feet per minute on some models, a flow camera has the advantage of being simple to operate. To the operator it looks very much like a rotary copier. Speed is usually fixed and after making any necessary adjustments to the strength of light in the exposure unit all that is necessary is to feed the documents into the copy slot.

Some cameras of this type are available with automatic document feeds while others are designed for use with continuous stationery. By exchanging the single document feed table or automatic feed for a continuous stationery feed it is possible on some models to film either continuous stationery or single documents.

Flow cameras always film on to roll film, never on to fiche. Although it is possible that one day someone will design a flow camera which will film on to fiche the technical difficulties are very considerable, to say the least.

M

FRAME   Each separate image recorded on a length of roll film or fiche is known as a frame. A double frame is two frames used to record an oversized original.

FRAME MARGIN   The unimaged area between the image and the frame.

GENERATIONS (IMAGES)   An image produced directly by the camera is said to be a first generation image; one produced from a first generation image is said to be a second generation image and so on.

GRID GAUGE   An inspection device used to determine whether the images have been correctly positioned on a microfiche.

IMAGE   The area within a frame which contains the representation of the original. The image may not fill the frame and will not do so where the format of the original differs from that of the frame. It is worth noting that the format of the ISO A paper sizes is the same as that of a standard microfilm frame, whether 16 or 35 mm film is used.

JACKET   A special kind of transparent carrier in which the microfilm is mounted in strips. A jacket normally has the same format as a fiche, i.e. several rows of film each of ten or twelve frames. The advantage of jacketed film is that the information can be updated by adding extra frames to an incomplete jacket or even removing some of the original frames and substituting new ones. Originally the film had to be mounted in the jacket by hand but a mechanical unit is now available for doing this automatically.

KALVAR FILM   A type of heat development film marketed by the Kalvar Corporation.

MICROFILM, SIZES   The most commonly used size for filming business documents is 16 mm, exactly the same size as 16 mm movies. Eight millimetres is also used sometimes although the range of equipment available in this size is limited. Film of 35 mm is used for filming engineering drawings and can also be used for oversized business documents, while 70 mm film is used for filming over-sized plans, especially by the construction industry.

Generally speaking 16 mm microfilm is suitable for recording all documents up to A4 or 13 by 9 in but much depends both on the quality of the camera and on the image on the original. Other things being equal it is possible to film at a greater reduction on a planetary than a flow camera and an original on which the image is in bold well-spaced type will stand greater reduction than one which contains fine print. According to BS 4187 the reduction ratio for filming business documents on to fiche should not exceed 1:22 but this is exceeded on some equipment as is the maximum reduction of 1:30 recommended for engineering drawings in BS 4210. In the case of large reductions only trial and error will show whether the originals are suitable.

MONOBATH, FILM   A type of film which requires only one chemical bath to process. It is simpler to process than normal film and some equipment is provided with a lightproof cassette so that a short length of monobath film can be guillotined from the roll and developed on the spot.

NEGATIVE   A negative is an image having clear lines against an opaque (black) background.

ORIGINAL   Any document (or drawing) which is to be micro-filmed is referred to as the original.

PCMI   A proprietary system by NCR enabling 3,010 images to be recorded on a 4 by 6 in fiche, a linear reduction of 150:1. This extremely small image needs special equipment both to film the image and to retrieve it.

PLANETARY CAMERA   This refers to a type of camera on which the lens is mounted directly above the copyboard, assuming the camera is of the vertical type. Each original is mounted separately on the copyboard and the camera focused exactly on to the document to be copied. Originally adjustments as to exposure, focus, etc. were made manually but there is an increasing tendency for these to be done automatically, which makes them quicker and easier to operate.

Several types of planetary camera are available, some being designed to film engineering drawings on to 35 mm roll film, while others are designed to film directly on to fiche. These latter have a mechanism which ensures that the lens is exactly lined up with the document regardless of the frame on the fiche being filmed.

POSITIVE   An image having opaque lines against a clear background is a positive, the opposite to a negative. It should be, noted however, that whereas in general photography a negative is usually an intermediate, in microphotography it is frequently used as the end product.

READER   A reader is a device which enlarges the image on a frame of microfilm and throws it on to a screen so that it can be read. In principle it is the same as a viewer used by amateur photographers to look at 35 mm transparencies. Most modern readers use the back projection principle to throw the image on to a translucent screen built into the reader. There are, however, some inexpensive readers which have no screen and throw the image on to any handy surface such as a wall or desk top. Some readers are made to accommodate fiche, while others are made for roll film. Generally speaking fiche readers are less expensive than the roll film type because the mechanism needed to transport fiche from one frame to the next is less expensive to make than a roll film transport. In any case winding roll film by hand is tedious so most roll film readers are powered and designed to take cassetted film. Many are also provided with devices for locating any particular frame on the roll. The simplest of these is a counter which when used in conjunction with a counter on the camera and an index enables any desired frame to be identified quickly. Other more elaborate devices designed for very large volume files incorporate optical coding connected to the transport in such a way that a frame can be selected by pushing buttons on a keyboard. It is also possible to get fiche readers on which any frame can be selected automatically under push button control.

READER/PRINTER   In addition to throwing an image on a screen a reader/printer is also capable of producing an enlarged hard copy. Depending on the model a number of different processes are used. These include most of the copying processing for which projection speed materials are available. However, as with copying the chemical processes are being rapidly replaced by the electro-static method. This enables prints to be produced quickly with virtually no skill on the part of the operator.

ROLL FILM   Roll film as compared to fiche is used where a large volume of originals need to be filmed, and filed. A standard 100 ft roll of 16 mm film contains 2,400 images but once the originals have been filmed it is possible to cut the film into strips and mount it on fiche or in jackets. To some extent the quality is bound to be lost where this is done. Nevertheless the standard is still good enough for the majority of business applications. Whether the film is retained in roll film and retrieved in that form or whether one of the other systems is used must depend on the size of the file and on the application.

Index